Coal Liquefaction Fundamentals

Coal Liquefaction Fundamentals

D. Duayne Whitehurst, EDITOR

Mobil Research and Development Corporation

Based on symposia sponsored

by the Division of Fuel Chemistry

at the ACS/CSJ Chemical Congress,

Honolulu, Hawaii, April 2–5, 1979

and at the 178th Meeting of the

American Chemical Society,

Washington, D.C.,

September 10–14, 1979.

A C S   S Y M P O S I U M   S E R I E S **139**

AMERICAN CHEMICAL SOCIETY

WASHINGTON, D. C.   1980

Library of Congress CIP Data

Coal liquefaction fundamentals.
 (ACS symposium series; 139 ISSN 0097–6156)

 Includes bibliographies and index.

 1. Coal liquefaction—Congresses.
 I. Whitehurst, Darrell Duayne, 1938–   . II. Ameri-
can Chemical Society. Division of Fuel Chemistry. III.
ACS/CSJ Chemical Congress, Honolulu, 1979. IV.
Series: American Chemical Society. ACS symposium
series; 139.

TP352.C64   662'.6622   80–20585
ISBN 0–8412–0587–6   ACSMC8   139 1–411 1980

PRINTED IN THE UNITED STATES OF AMERICA

ACS Symposium Series

M. Joan Comstock, *Series Editor*

FOREWORD

The ACS SYMPOSIUM SERIES was founded in 1974 to provide a medium for publishing symposia quickly in book form. The format of the Series parallels that of the continuing ADVANCES IN CHEMISTRY SERIES except that in order to save time the papers are not typeset but are reproduced as they are submitted by the authors in camera-ready form. Papers are reviewed under the supervision of the Editors with the assistance of the Series Advisory Board and are selected to maintain the integrity of the symposia; however, verbatim reproductions of previously published papers are not accepted. Both reviews and reports of research are acceptable since symposia may embrace both types of presentation.

CONTENTS

LIQUEFACTION MECHANISMS

PREFACE

This book is composed predominantly of a compilation of papers presented in two symposia of the Fuel Division of the American Chemical Society. These symposia were held originally at the joint ACS/CSJ Chemical Congress in Honolulu, Hawaii, session on coal liquefaction fundamentals, April 1979, and the 178th Meeting of the American Chemical Society, at the Storch Award Symposium in Washington, D.C., September 1979.

One purpose of the former symposium was to bring together scientists from different parts of the world to discuss their beliefs on the critical features of coals found in their countries that are relevant to liquefaction behavior. The first chapter provides an excellent general background on the biological and geological origin of coals from different parts of the world and the significance of this to liquefaction potential.

In the rest of this section, coal liquefaction researchers of the United States, South Africa, Australia, Japan, Canada, and Great Britain describe their country's coal reserves and origins, and the significance of coal composition to liquefaction behavior.

The second section of this volume describes several potentially new liquefaction processes which may have higher efficiencies than today's developing technologies. The theme of the Storch Award Symposium, featured throughout these six chapters, was new process potentials through the use of short-contact-time thermal processes followed by catalytic upgrading.

The mechanisms by which coal is converted to soluble or liquid form and the nature of the products of such reactions have been the subjects of a great deal of effort throughout the world. In the last two sections, researchers from Australia, Japan, South Africa, and the United States describe their findings in these areas. The reader will note that no unanimous agreement exists on the chemical mode by which coal is converted although kinetic descriptions are often similar.

This book is intended primarily for those who have some expertise in coal liquefaction but the first six chapters should be very valuable to persons just entering the field. The international flavor of the text is some-what unique and will provide the readers with a feeling for goals and philosophies for coal liquefaction in various parts of the world.

Mobile Research and Development Corporation
P.O. Box 1025
Princeton, NJ 08540

May 21, 1980

D. DUAYNE WHITEHURST

RELATIONSHIPS BETWEEN COAL COMPOSITIONS AND LIQUEFACTION BEHAVIOR

Some Proved and Unproved Effects of Coal Geochemistry on Liquefaction Behavior with Emphasis on U.S. Coals

P. H. GIVEN, W. SPACKMAN, A. DAVIS, and R. G. JENKINS

College of Earth and Mineral Sciences, The Pennsylvania State University, University Park, PA 16802

1. Introductory Remarks on Coal Paleobotany, Geology and Geochemistry

The purpose of the collection of papers in this volume is to review what can be said about the susceptibility to lique-faction of coals from different parts of the world. We and later authors will present data relative to coals of the areas with which we are familiar. However, with the Editor's approval, we are going to devote the first part of this paper to making some general remarks about coal geology and geochemistry, in the hope that this will provide a useful background to what comes later.

1.1 Paleobotany of Coal Origins

On the evidence of coalified organs and tissues of the higher plants identifiable under the microscope in some coals (1), or petrified plan tissue found as "coal balls" in some coals (2), and on the basis of the pollen and spore content characteristic of coals, it is generally accepted that they are derived mostly from the organic matter of the higher plants, altered to a greater or lesser extent by microorganisms (3,4), and partly in some cases from the lower plants (algae).

The hydrologic, geologic and climatic conditions necessary for the formation of extensive coal measures are evidently rather specific, because the periods in geologic time in which major epi-sodes of coal formation occurred were sporadic in any one geogra-phic area (see Table I). Thus is many areas, there have been large gaps in time when no major coal measures were formed. Since the evolution of the plant kingdom proceeded continuously, whether or not coal measures were being laid down, the coals that formed after a gap in time were formed from plants quite different to those that gave rise to coals before the gap. Table II summarizes a few important events in the evolution of the plant kingdom.

0–8412–0587–6/80/47–139–003$08.00/0

Table I. Major Eras of Coal Formation

	North America	Europe	Far East	Southern Hemisphere	in m. years B.P. approximately
Cenozoic					
Pliocene	+ (Alaska)	+	—	—	10
Miocene	—	++	—	+ (Australia)	20
Eocene	++	+	++	—	45
Mesozoic					
Cretaceous	++	—	++ (Japan)	++	100
Jurassic	—	—	++	++ (Australia)	165
Triassic	—	+	—	++ (Australia)	200
Paleozoic					
Permian	—	—	—	++ (All Gondwanaland)	250
Carboniferous	++	++	—	—	310

++ coals very abundant
+ abundant
— absent

Table II. <u>An Outline of Plant Evolution</u>

Period	Approx. m. years before present	
Late Silurian	400	first appearance of lignified land plants
Carboniferous	350–270	large, diverse flora, of spore-dispersing plants, including ferns and slender trees with varying amounts of branching and leaf development
Permian	270–225	seed-fern flora (Glossopteris) flourishes all over Gondwana-land
Triassic & Jurassic	225–180	seed-bearing plants flourish with conifers and cycadophytes prominent
Cretaceous	135– 70	flowering plants evolved (Angiosperms)
Late Cretaceous	say 80	essentially modern flora

A comparison of Tables I and II shows that the major coal measures of the world were derived from several quite different floristic assemblages. Thus the plants that gave rise to coals in Europe and the east and midwest of North America in the Carboniferous were part of a complex flora that included ferns, seed ferns, horsetails, lycopods and conifer precursors. Except for the conifer precursors, lignified xylem tissues tended to be minimal in these plant groups and unusually large leaves with extensive waxy cuticles were characteristic of three of the groups. The abundant lycopods emphasized the development of a "corky" periderm or bark which, presumably, was formed of cell walls that were heavily impregnated with suberin as well as lignin. (Both cutin and suberin appear to be polymers of long chain hydroxy acids. Suberin is a waxy substance developed in the thickened cell walls characteristic of cork tissues.) Most of the plants were prolific spore-formers and generated large quantities of thick-walled microspores and megaspores whose waxy exines proved particularly resistant to decay and decomposition. Finally, the ratio of purely cellulosic cell walls to walls impregnated with lignin or suberin was much higher than at any subsequent time.

The term "Gondwanaland" in Table I refers to a once-existing supercontinent consisting of what are now known as Africa, South America, Antarctica, Australasia and India (5). The Permian coals

of this supercontinent in the southern hemisphere were all formed
under climatic conditions differing from those that prevailed in
the northern hemisphere during the Carboniferous, with important
consequences for the petrographic make-up of Gondwanaland coals (5).
Also, these coals were formed from a less diverse flora than that
described above, with a particular group of seed ferns and pre-
conifers playing a conspicuous role in the vegetation. The seed
habit began to affect the volume and type of spore production and
the ratio of lignified to non-lignified tissues probably increased.
Lycopods with their suberin-rich "barks" were no longer significant components of the swamp floras.

The Little paleobotanic study has been made of the coal-forming
plants of the Triassic and Jurassic in Australia and Africa, but
in the rest of the world two plant groups ascended to positions
of dominance. These were the conifers and the cycadophytes. The
cycadophytes emphasized non-lignified parenchymatous tissue (i.e.,
containing living, protoplasmic cells) and large fern-like leaves
with well developed waxy cuticles. The conifers emphasized the
development of massive lignified stem and root cylinders with a
significant amount of suberized periderm. Most of the plants in
both groups were of arboreal habit, although the cycadophytes were
often short and stocky, with much of their mechanical support as
engineering structures being provided by heavily lignified leaf
bases. All were seed producers, meaning that they no longer formed large quantities of the thick-walled megaspores that are so
characteristic of many Carboniferous coal types.

The Angiosperms evolved in the Cretaceous, and from that
period on the coals were formed from floras much like those we see
today. In any period, the floral origin of the coals in one basin
was not necessarily precisely the same as that of coals formed in
other basins at the same time.

The above outline of the evolution of the plant kingdom during coal-forming eras has been presented largely in the language
of botanical anatomy. However, the alert chemist will note that
the anatomical differences imply considerable quantitative and
spatial differences in the distribution of the principal plant
constituents [cellulose, lignin, cutin, suberin and other waxes,
contents of protoplasmic cells, pigments, resins, sporopollenin.
The latter substance is thought to be the principal constituent
of the outer layer (exine) of both spores and pollen grains; it
is said to be a co-polymer of oxygenated carotenoid compounds
with long-chain fatty acids (6,7)]. The differences may be qualitative also. It is a prime characteristic of the higher plants,
as opposed to other types of organism, that they contain a wide
variety of phenolic substances, of which lignin is only one (8,9).
Other important phenolic constituents of plants include the hy-
drolyzable and condensed tannins (9) which as cell fillings are
thought to give rise to the rather rare coal maceral, phlobaphen-
ite (10).The structure and nature of these various phenolic substances differ considerably in plants that represent differing

degrees of evolution (11,12,13,14). Thus the "paleo-biochemistry" of coal-forming plants in different periods and areas differed in several respects.

1.2 Metamorphism or Catagenesis

There are geologic processes by which a peat formed at the earth's surface becomes buried progressively more and more deeply, and it is chiefly exposure to the elevated temperatures experienced at depth that is responsible for converting peat into the coals of various ranks, the rank attained depending on the maximum temperature reached by a stratum (15,16). The mean temperature gradient in the earth's crust is $3°C/100$ m, but there is wide variation about the mean. Also, the length of time during which deeper and deeper burial takes place, before uplift brings the stratum back towards the surface, is quite variable. Thus coals in different basins may reach the same apparent rank through materially different temperature/time histories (as, for example, in the Appalachian and midwest areas of the U.S.)(16). The metamorphism of coals no doubt involves a large number of parallel reactions, each having its own enthalpy and entropy of activation. Available methods of assigning a degree of metamorphism, or rank, to a coal are so crude that it is very likely that we shall describe two coals as of the same rank when, even if the starting materials were identical, a different mix of products results from coal A being formed at temperature T_1 during the time interval t_1, while coal B was formed at the lower temperature T_2 during the longer time t_2. For pure substances there is in kinetics a considerable degree of temperature/time compensation: for coals, this cannot be so, because of their complexity.

It is customary to treat the kinetics of processes altering coals by the classical methods of chemical kinetics, as if a coal were a single chemical substance, whether from the point of view of the geochemistry of metamorphism (17), or laboratory pyrolysis (e.g.18), or liquefaction (e.g.19). In a study of coal liquefaction mechanisms and kinetics, Szladow has strongly objected to such procedures, arguing that any valid kinetic analysis must start from the proposition that any chemical process of alteration of a coal consists of a large number of reactions proceeding in parallel, each with its own rate and temperature coefficient (20,21). Accordingly, he developed his own kinetic analysis of liquefaction, which showed, *inter alia*, that the apparent overall energy of activation of coal liquefaction must vary with the degree of conversion of the coal to liquids + gases. Surely similar considerations must apply to the complex of chemical reactions that represent metamorphism, or, as the low-temperature thermal alteration of organic matter is increasingly being termed, catagenesis (22).

The various aspects of coal origins briefly reviewed above already strongly suggest that a world view of the interrelationships of coal properties, and of the relation of coal characteristics to behavior in processes, is likely to show a great deal of dispersion. Very few authors have obtained data that confirm or deny this conclusion. Tribute should be paid to Mott(23), who, 37 years ago, showed that the moisture-holding capacity of some 2000 coals is dependent on geological antecedents as well as on rank.

1.3 Some Macerals and their Origins

Certain compositional differences between coals of differing origins can be inferred from available data. Differing anatomical distributions of cellulose, lignin and suberin, with implications for the origins of vitrinites, and differing distribution of phenolic substances in plants of different orders and families, have been referred to above. Some biochemical investigations of modern representatives of ancient plants have been made (e.g., refs. 14, 24), which display taxonomic variations in lignin structure, flavonoid types and sugar anabolism.

Most of the plants that flourished in the Carboniferous and Permian reproduced by means of spores. This was a rather inefficient mode of reproduction, so that large numbers of spores were produced as a fail-safe reproductive strategy. Later plants (conifers and flowering plants) reproduced more efficiently via seed production. This involves elimination of the development and dispersal of large quantities of thick-walled megaspores and the production of smaller quantities of, often, thinner-walled pollen grains as the equivalent of the more ancient microspores. Thus the coals formed in the Triassic and later typically contain considerably smaller concentrations of sporinite (or exinite) maceral than earlier coals, and in biochemical terms the contribution of sporopollenin was therefore much less. Since it is assumed that sporinite is a highly reactive maceral in liquefaction, this is significant.

So far as is known, all of the higher plants contain terpenoid hydrocarbons and oxygen-containing compounds that are commonly described as resins. These may have various biological functions, such as sealing of wounds in the stem by exudation of liquid terpenoids and oxidative polymerization to a solid resin. Resin ducts containing such materials did occur in the plants that gave rise to coals in the Carboniferous, but were much more abundant in the conifers that were the precursors of many coals in the Cretaceous. Thus the distribution of resinite macerals differs in coals originating in different eras. The study of Murchison (25) suggests materially different chemical characteristics of resinites of differing geological age. Fragmentary unpublished observations of our own indicate that resinites are highly reactive macerals in liquefaction.

Fusinite macerals are generally held to be inert in coking, and in liquefaction, as will be seen below. In the production of metallurgical coke, they do not become fluid on heating, and participate in the formation of vesicular coke only as non-reactive diluents (though contributing to the mechanical strength of the coke). Such knowledge as we have of fusinites is entirely based on samples procured from macroscopically visible lenses of fusain that often occur in coal seams. Fusinites of this kind have long been believed to have originated as charred wood formed in forest fires (26). Comparisons of the change in e.s.r. signal given by pairs of vitrinites and fusinites from the same coals after laboratory pyrolysis showed that the fusinites had already been exposed to temperatures of 500-600°C (27,28). From this point of view, related semifusinites must represent the inner layers of woody stem that were partially protected and experienced relatively low temperature pyrolysis; hence they are likely to show a range of compositions and reactivities.

The kind of fusinite discussed above is called "pyrofusinite" (26). It is of high reflectance (mean value, 3-5%) and often preserves the cellular structure of wood (26); it is of high carbon content (29), and high aromaticity (29,30,31). Thus there is some degree of understanding of the origins and nature of pyrofusinite.

There is, regrettably, little understanding of the nature or properties of another important sub-maceral of fusinite, "degradofusinite". M. Teichmüller believes that this maceral may originate in the unconsumed part of wood attacked by fungi such as dry rot, *Merulius lachrymans*; such fungi "alter the unused part of the wood into carbon-rich, strongly reflecting, humic substances" (26). For what reason or by what means the fungus should so alter "the unused part of the wood" is not made clear. Moreover, no evidence that degradofusinite is indeed carbon-rich has ever been presented, that we know of, and in fact we have experience of only one single concentrate of this maceral, which was prepared by M. Teichmüller; unfortunately, its carbon content was not determined, but its e.s.r. behavior indicated that it had not previously been exposed to elevated temperatures (32). In any case, the proposition that some microorganisms in their respiratory cycles or other metabolic processes burn the hydrogen from organic molecules to leave a highly carbonaceous residue postulates physiological activities that have not so far been recognized in any laboratory culture or natural ecosystem that we have heard of. Of course, it is plausible to suppose that there may be biochemical processes that transform plant tissue into the precursor of degradofusinite, if the assumption were removed that this maceral is highly carbonaceous; we are, of course, ignorant of what these processes may be.

Concerning the distribution of degradofusinites, M. Teichmüller states (26): "By far the greatest part of Carboniferous and Permian fusinite and semifusinite occurs in dull coal bands (durains). These fusinites, in contrast to the fusain lenses, are not visible macroscopically. Their cell structures are poorly preserved, their reflectivity is mostly semifusinitic......these degradofusinites......apparently form through dehydration and oxidation", and again, "Degradofusinites are especially abundant

in the thick Carboniferous bituminous coals of E. Upper Silesia
[Poland] and in many Gondwana coals which may contain up to 50% of
this material".

There are certainly lithotypes that can be handpicked from
European and American coals that are relatively rich in fusinite
and semifusinite. However, it is perhaps significant that the
mean content of *total* fusinite + semifusinite in 697 coal samples
in the Penn State/DOE Data Base is 8.9%. On the other hand, the
content of inertinite macerals in the Permian coals of Gondwanaland is notoriously high and much of this inertinite material consists of semifusinite (5,26,33,34), the concentration of which can
be as high as 50% in the whole seam.

Gray et al. (33) state that one third of the semifusinite in
Permian coals of South Africa is assumed to be reactive in coking;
this is the same proportion as is assumed reactive in U.S. coals
by some workers, as will be documented later. In fact, it seems
to be assumed that pyrofusinite and degradofusinite have the same
reactivities (35). However, the evidence for this proposition is
scanty and it is evidently undergoing a careful re-examination,
as is made clear in the contribution to this symposium by Durie (34).

Thus we accept that the maceral "degradofusinite" and the
associated semifusinite exist, that they are widely distributed
and quantitatively important in many Gondwanaland coals, and that
their origin is different from that of pyrofusinite. However, for
biochemical reasons, we are unable to accept the suggestion that
these macerals are derived from wood unconsumed by fungi if they
are indeed highly carbonaceous, and we submit that the character
and technologic properties of some major constituents of coals of
the southern hemisphere including India, are hardly understood at
all.

The recent presentation of Neavel draws attention to the
heterogeneity of coals and its importance in determining the properties and conversion behavior of coals (36). The remarks made
above indicate that the petrographic composition of coals varies
with the character of the flora from which their organic matter
was derived and that the nature of some of the macerals or sub-
macerals, as presently judged, may be different in different areas.
The data reviewed by Francis (37) show that coals from, for example, Nigeria and Pakistan may contain particularly high contents
of hydrogen. Petrographic analyses by modern methods were not
given, but the data presumably testify to the presence of relatively high concentrations of macerals of the liptinite suite (sporinite, cutinite, resinite, alginite). Thus if one takes a global
view, it appears that there can be wide variations in petrographic
composition and hence in liquefaction behavior, since macerals of
the liptinite suite, of highly aliphatic character, are assumed
to be specially susceptible to liquefaction.

1.4 Further Aspects of the Effects of the Environments of Deposition and Coalification

At the present time, a large majority of the peats now accumulating are to be found in fresh-water hydrologic conditions (4, 38), whereas in the geologic past, saline conditions often prevailed (4,38,39). Sulfate ion is the second most abundant anion in saline waters, and whenever such waters impregnate an accumulating organic-rich sediment, bacterial reduction of sulfate to H_2S, and fixation of sulfur as sulfide minerals and in organic structures, occurs (4,40,41,42). This is obviously an important factor determining the sulfur content of coals, which, as we shall see, is in its turn an important factor determining liquefaction behavior. In addition, one wonders whether the combination of halophilic plants as source of the debris preserved in peats with the activities of halophilic microorganisms, and the reducing conditions resulting from the abundant presence of H_2S, might not result in the preservation of organic matter different in some ways from that accumulating in fresh-water conditions. The degree of salinity, and hence the magnitude of these effects, is likely to vary in different sites of accumulation (43).

Teichmüller (39) has indicated that coals formed in saline environments tend to be richer in hydrogen and nitrogen than fresh-water coals. She also believes that certain fluorescent macerals may be relatively more abundant in coals formed in more saline conditions; in accordance with this view, fluorinite and fluorescent vitrinite appear to be more abundant in coals from Illinois than in those in the Eastern province.

Recent work has suggested that the coals of the Illinois Basin were never buried deeper than about 1500 m. (44), compared with an estimated 3000 m. or more for the coals of western Pennsylvania in the Eastern province. Presumably as a consequence, the coals of the Interior province tend to show low values of vitrinite reflectance and high values of moisture-holding capacity relative to coals of other areas of apparently similar rank (45).

The minerals and other inorganic species in coals derive mostly from the input of detrital minerals and soluble cations to the original peat swamp or marsh; these in turn derive from whatever rocks are being eroded in highlands around the peat-forming area. This input makes possible the formation of additional minerals within the peat (e.g., pyrite, calcite). Also, some further mineral deposition (e.g., of pyrite) may occur in cleats after induration of the organic sediment, to an extent that again depends on local geochemical conditions (46,47).

1.5 Some Conclusions of this Section

All of the material outlined in this introductory section suggests strongly that the coals of the world almost certainly will show a wide variety of characteristics, due to differing

plant origins, to differential preservation and alteration of
plant tissues and organs, to differing inputs of inorganic materials and to differing temperature/pressure/time histories. Surely, therefore, one must anticipate considerable dispersion of
behavior in conversion processes. However, it is doubtful whether
any institution has either the financial resources or a sample
base adequate to permit a test of these hypotheses.

A further factor is that in any one continental land mass,
at any one point in time, there have been a number of different
environments of deposition of peat, which may well have influenced the characteristics of the coals derived from the peats. It
is not proposed to review these in any general sense, but some
discussion is offered later on the possible relevance of this factor in determining the liquefaction behavior of U.S. coals. In
the meantime, we summarize, for future reference, the major coal-
bearing areas, or provinces, of the United States in Table III.
The relevance of the remarks on sulfur content is that, for reasons explained above, it is usually a valid index of the salinity
of the environments of deposition. It was remarked earlier that
the Eastern and Interior provinces have experienced different
temperature/pressure/time histories. It should be added that
coals of the Rocky Mountain, Pacific and Alaskan provinces most
probably experienced yet further sets of conditions of metamorphism: a locally increased geothermal gradient that produced
relatively high temperatures at relatively low depths of burial
and hence at relatively low pressures of overburden.

2.   Experimental Studies of the Dependence of Liquefaction Behavior of U.S. Coals on Coal Characteristics

The remainder of this paper can be restricted to coals of
North America, since this is the area for which we have data and
in any case other contributors to this collection will deal with
the coals of their own areas. The first statement above needs
qualification: we ourselves have no liquefaction data on Canadian
coals, but Ignasiak et al. (48) present some in this collection.Re-
lying, as in the earlier part of this paper, on geological information, we can say that the strata of the North Great Plains and
Rocky Mountain provinces continue north into Canada, as does the
Pacific province. Nova Scotia contains some Carboniferous coals
related to those in the Eastern province.

2.1   Early U.S.B.M. Work

A large program of work on coal liquefaction at the U.S.
Bureau of Mines station at Bruceton, Pa., under the direction of
H. H. Storch, was stimulated by the pre-war and wartime develop-
ments in Germany (49,50,51,52,53). The very extensive studies
showed that, with some modification of processing conditions, most
U.S. coals could be converted to liquid fuels in acceptable yields

Table III. The Coal Provinces of the U. S.

Province	Age	Geographic Area	Range of Rank	Other Comments
Eastern (Appalachian)	Carboniferous	Pa., Ohio, W. Va., Ala., Tenn., E. Ky.	HVB to anthracite	variable S (medium to high)
Interior (E. region)	Carboniferous	Ill., Ind., W. Ky.	HVC to HVA	high S
(W. region)	Carboniferous	Kan., Okla., Iowa, Ark., Tex.	HVC to HVA	high S
North Great Plains	Cenozoic	N. and S. Dakota, parts of Mont., Wyo., Ariz. and N. Mex.	Lignite to HVB	low S, very large reserves of lignite
Rocky Mountain	Cretaceous	Utah, Colo., parts of Wyo. and N. Mex.	sbb to HVA	low S, many separate basins
Pacific	Cenozoic	Wash. (Calif.)	sbb to HVA	low S, much influenced igneous intrusions at depth
Gulf	Cenozoic	Tex., parts of Miss., Ala., Lou.	Lignite	medium S
Alaskan	Cenozoic	Alaska	sbb to HVA	low S

See Table 1 for approximate absolute ages of the periods stated.

(the severe conditions needed for some coals would now be exces-
sively expensive). Detailed studies were made to evaluate the
liquefaction potential of petrographic constituents of coals (50,
53). The system of petrographic analysis used is now regarded as
obsolete, and entities described as "translucent attritus" and
"opaque attritus" cannot be interpreted unambiguously in modern
terms (contrary to some unsupported and ill-advised statements in
the literature, 54,55). However, the findings that all "fusains"
tested showed poor liquefaction conversion is acceptable evidence
that fusinite macerals are unreactive in liquefaction. An important review of this and related work was published in 1968 by Wu
and Storch and has recently been reprinted (54).

2.2 Influence of Rank

Coal rank is the most obvious characteristic to examine in
relation to variations in liquefaction behavior. Different researchers have reached varying conclusions on the effect of rank in
coal liquefaction experiments (e.g. 50,56), although there is general agreement about the lower yields obtained from higher rank
coals. Some have reported that the highest yields are obtained
from the lowest rank coals, others that optimum conversion is
found for coals of high volatile bituminous rank, and yet others
that there is no satisfactory rank-conversion relationship. Workers at the U. S. Bureau of Mines (50) concluded that low-rank samples were more sensitive to experimental conditions than bituminous
samples. More recent results from different laboratories also lack
consistent trends with rank. Our own limited data obtained from
two series of autoclave experiments (57) indicated that maximum
conversion occurs at high volatile bituminous rank.

Contrasting results were experienced with a suite of seven
coals, all of high vitrinite content and varying in rank from lignite to low volatile bituminous, reacted in Mobil Research and
Development Corporation's stirred autoclave for a very short residence time and with a synthetic solvent (58) (430°C, ᴧ13.40 MPa
of H_2, 3 min. residence time, coal particle size <350μm). These
experiments were designed anticipating that a shorter residence time
would be more likely to reveal the effects of differences in the
conversion mechanisms of the individual coals. Figure 1 shows
the conversions obtained from this suite, the greatest being for a
high volatile A bituminous coal (vitrinite $\bar{R}_{max}$ = 1.00%). The
Mobil R&D group have suggested that the rate of dissolution of the
lower rank coals is sufficiently slow to be reflected in the conversion levels achieved at 3 min. residence times (59).

Further information on the trends of behavior of vitrinites
with change of rank was provided by the examination with an optical
microscope of residues from liquefaction experiments. This will be
dealt with later.

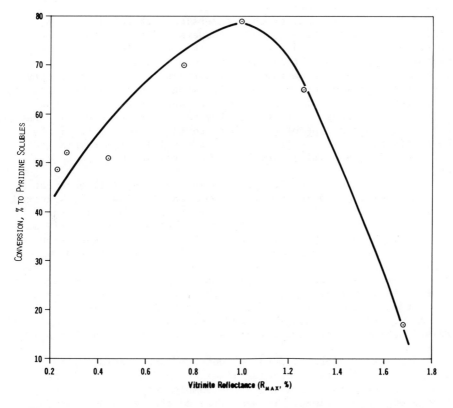

Figure 1. Conversions over short residence time for high-vitrinite coals of varying rank

2.3 Statistical Analyses of Liquefaction Data for a Large Sample Set

The above discussion has assumed that the rank of a coal can be adequately measured by a single parameter, such as the reflectance, the volatile matter yield or the organic carbon content. This assumption is commonly made, but it has for a long time appeared a pretty improbable proposition. The discussion also was restricted to bivariate correlations, that is, plots of a single variable against another.

A statistical study of the conversion with tetralin of 68 coals (60) must now be regarded as superseded by a later, more comprehensive paper (61), but it did show very clearly that bivariate plots are of little value in interrelating liquefaction behavior with coal properties; at least two or three coal properties must be taken into account in seeking to explain the variance of liquefaction behavior, and some of these properties are not related to the rank of the coal. The paper implies strongly that any interrelationships of coal characteristics must necessarily be multivariate. Hence in any study of coal a large sample and data base is essential if worthwhile generalizations are to be made.

The raw data in the more comprehensive study (61) were conversions, determined in duplicate, when each of 104 coals selected from three geological provinces was heated with tetralin under standard conditions, together with the results of 14 commonly made analytical determinations for each coal. An early observation in this study was that when data for all 104 samples were plotted against volatile matter, a steady decrease of conversion with decreasing volatile matter was apparent. But there was a great deal of scatter (r=0.85). In any case, the formal requirements that make possible the employment of valid statistical analyses were not met by the data matrix, as evidenced by skewed and bimodal relationships between the variables: the sample set was heterogeneous.

"Cluster analysis", when the computer was provided with 15 coal characteristics, including liquefaction conversion in reaction with tetralin, partitioned the sample set into three more homogeneous populations. It was observed that each population consisted mainly of coals from one particular geological province, and, interestingly, the groups differed markedly in total sulfur content and liquefaction conversion, and differed significantly in rank as measured by the carbon content: discriminant analysis showed that the contents of total sulfur and of organic carbon could by themselves be used to assign correctly 102 out of 104 coals to the groups revealed by cluster analysis.

Principal components analysis of the characteristics of the coals in each of the three groups showed that the interrelationships between coal properties were markedly different; that is, the trends of properties with increasing rank are different, in

important ways, for coals assigned by cluster analysis to different groups. Moreover, multiple regression analyses, designed to
establish predictive relationships between liquefaction conversion
and coal characteristics, called out a different selection of coal
properties for each of the three groups. The geochemical basis
of these empirical findings is something that vitally needs to be
understood as a contribution to the advance of coal science and
technology. However, at least they demonstrate (albeit empirically) that the geological/geochemical antecedents of a coal are relevant to its properties and behavior, and provide clues indicating
where searches for scientific understanding should start. The
provinces from which coals were selected were the Eastern (or
Appalachian), the Interior (the area known to Americans as "the
mid-west"), and the Rocky Mountain. The first two are of Carboniferous age, and the third of Cretaceous age, so that the character of the antecedent vegetation will differ as described earlier.
Moreover, the temperature/pressure/time histories of the strata
differed markedly, as has been briefly described in a recent publication (62), which also provides evidence that there are system-
atic structural differences between coals of the different provinces.

About one half of the coal samples used in the above study
(61) have been investigated by workers in Gulf Research and Development Company, using a continuous flow reactor (63). The
throughput was about 1 kg./h of coal/solvent slurry, the solvent
was a partly hydrogenated anthracene oil, temperatures of 440 and
455°C were used, and the system was pressurized with hydrogen to
20.69 MPa.

If the coals studied are classified into the groups established by the cluster analysis discussed above, Gulf conversion
data for coals of groups 1 and 2 are even more widely segregated
than they are for the behavior of the samples in our small-scale
batch reactors, as illustrated in Figure 2. [Group 1: coals of
relatively high bituminous rank, of medium sulfur content, very
largely from the Appalachian province. Group 2: coals of the
medium and low ranges of the high volatile bituminous classes, of
high sulfur content, mostly from the Interior province, but with
a substantial minority from the Appalachian province (61). Group
3 coals had not been studied at the time the paper (63) was written; they are with only one exception from the Rocky Mountain
province.] Statistical analysis empirically shows good linear
correlations between our data for small batch reactors, and Gulf
data for a different donor solvent, the use of continuous flow
conditions and pressurization with 20.69 MPa of hydrogen (63).
Thus the predictive ability we have established from small-scale
batch tests most probably has relevance under conditions more
realistically representing commercial practice.

Considerable difficulty has been experienced in removing
residual solid organic matter and mineral constituents by filtration or centrifugation of the products issuing from coal liquefac-

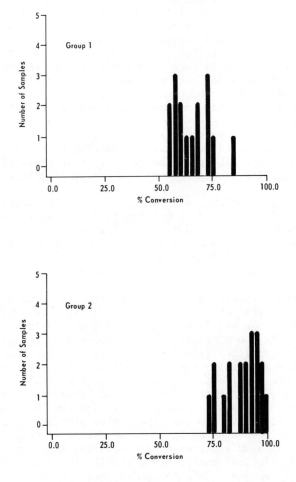

Figure 2. Distribution of liquefaction conversion for coals tested in Gulf continuous-flow unit (455°C)

tion pilot plants. Several processes now under development (notably SRC II and Exxon Donor Solvent) tend to rely on vacuum distillation as a means of removing solids from the saleable product. Hence, the distillability of products can be an important aspect of the liquefaction behavior of coals. Results obtained under the subcontract from P.S.U. to Gulf show that the fraction of coal feedstock converted to liquids distillable under a standard set of conditions varies widely from coal to coal (e.g., from a low of 12% to a high of 60% of dmmf coal) (63). We have yet to identify the characteristics of a coal that make for a high yield of distillable products. A plot of yield of distillate against the carbon content of the coal showed a reasonably good correlation, distillability decreasing with increasing rank. (63) The coals of Group 1 tend to give a lower yield of distillable material than those of Group 2: because Group 1 coals tend to be of higher rank, or for some other reason?

It is known that pyrite, or the pyrrhotite derived from it during liquefaction of coals, catalyzes coal liquefaction, increasing conversion to liquids + gases and to liquids soluble in hexane, and decreasing the sulfur contents of the products (see, for example, Granoff, 64). We have observed (Yaykin and Given, 1979-80, unpublished) that pyrite specimens isolated from a number of different coals exhibit differing levels of catalytic activity in the liquefaction of coals of very low sulfur content. Evidently the processes that emplace pyrite in coals give rise to differing particle size distributions, or differing electrical/magnetic/catalytic effects determined by the nature and concentration of impurity atoms or other defects in the crystal lattice. Here we see yet a further role of the geochemistry of coal formation.

Thus our rather small set of samples from a few selected areas of the U.S.A. shows a dispersion of some aspects of liquefaction behavior that is evidently associated with differences in the geology and geochemistry of the sample. Still more would we expect many sets of complex interrelationships between coal characteristics to emerge had we had a sufficiently large world-wide sample base to work with.

2.4 Microscopic Examination of Residues from Liquefaction

Our optical examinations of liquefaction residues have given further insights into the different responses of coals of varying rank. The huminite or vitrinite macerals of lignites and subbituminous coals, as might be expected, generally do not become plastic during liquefaction. The residues derived from members of this maceral group show similar particle integrity to the feed coal, although dissolution may leave tattered skeletons of the original structures (65). At the high volatile C bituminous rank level, plastic behavior of the vitrinite becomes evident in the appearance of the residues, and at high volatile B bituminous and higher ranks the vitrinite has plasticized more completely. This

is evident mainly through the formation of vitroplast (66), a
vitrinite-derived material occurring as isotropic spheres and
rounded particles in which no trace remains of the original bed-
ding structure. Vitroplast, particularly that from high and medium
volatile bituminous coals, can develop pores owing to the formation of gas bubbles. At an early stage in liquefaction, the bloat-
ing up of vitroplast into highly porous cenospheres may contribute
to the dissolution of coal by providing a large surface area of an
intermediate plastic phase in which the molecular structure has
been loosened by depolymerization (67). However, if conditions
in the reactor are such as to permit a recombination of thermally
ruptured bonds, it is unlikely that the resulting cenospheres,
which have a high level of reflectance compared to either vitrinite or vitroplast, will be readily hydrogenated (66).

Some other results revealed by the rank series tested in the
Mobil autoclave referred to above are as follows:

a) Unreacted vitrinite was an important constituent only in
the residues from coals of high volatile C bituminous rank and
lower.

b) Significant amounts of anisotropic semicoke were formed
only by the low volatile bituminous coals.

The results of a) and b) together show that the higher rank
(medium and low volatile bituminous) coals do dissociate into a
plastic phase in the reactor. In the case of residues from low
volatile bituminous coals, these substances are readily coked,
as indicated by the development of mesophase spheres within the
isotropic pitch. Our work on solvent-refined coal (SRC) fractions
showed that the more highly functional, higher molecular weight
fractions have a greater propensity to form pyridine-insoluble
products. Further, the more pyridine-insoluble material that is
produced by heating various SRC's, the greater the amount of mesophase semicoke formed appears to be.

2.5 Behavior of Macerals other than Vitrinite

Earlier publications have documented the higher reactivities
of vitrinite and liptinite group macerals and the lower reactivities of certain inertinite macerals in liquefaction (50,57,68).

The degree to which fusinite and semifusinite react appears
to be related to the extent to which the woody tissues were altered by the processes leading to the formation of "pyrofusinite"
or "degradofusinite". As noted earlier, it is generally accepted
in the U.S. coke industry that fusinite, macrinite and micrinite
behave as inert substances in coke formation, whereas semifusinite
may behave partly as a reactive ingredient. In various coke
strength predictive methods employed by U.S. steel companies, an
arbitrary two-thirds of the semifusinite content is assigned to
the inert category (69). However, other workers have maintained
that semifusinites do not pass through a plastic phase (35). A
determination of how semifusinite behaves in liquefaction processes
is clearly important.

Table IV reports the maceral and reflectance analyses, and percent conversion in hydrogen transfer, for a series of samples high in inertinite macerals (70). The samples consist of two Australian coal seams with unusually high (by U.S. standards) semi-fusinite contents, and five hand-picked fusains of high fusinite and semifusinite contents from Illinois [whether the comparison of Australian and U.S. samples is justifiable is open to question, but the results are interesting]. The latter samples, presumably all of the pyrofusinite type, all show low conversions (12-25%), even where there is a relatively large percentage of semifusinite (in PSOC 263), and the mean fusinite-semifusinite reflectance is the lowest of the five Illinois fusains (3.05%). In contrast, the conversions obtained from the two Australian samples were moderate. PSOC 303 has a semifusinite content of 73%, so that it appears that this maceral is making a contribution to the total conversion of 40%. Note that the average reflectance of fusinite and semi-fusinite in this sample is only 1.26%. These various results suggest that there *are* differences between fusinitic macerals in Gondwanaland and northern hemisphere coals.

Microscopic examination of the residue of PSOC 303 confirms the liquefiability of the low-reflecting inertinite macerals. The remnants of semifusinite were observed which showed clear evidence of plasticity, including the rounding of particles and formation of spheres (66). In residues from the rank suite reacted in Mobil's stirred autoclave, semifusinite from U.S. coals of high volatile B bituminous rank and lower also appeared partly reacted, having rounded margins and dissolution embayments. That from higher rank coals showed no such evidence of plasticity, even though the high volatile A bituminous coal gave the highest level of conversion.

The behavior of macrinite and micrinite in industrial processes is not clearly understood. As stated above, many U.S. petrographers treat both of these constituents as "inert" coal constituents. On the other hand, overseas workers have observed that micrinite may not be inert during carbonization. Because some micrinite appears to have been generated during the progressive coalification of the liptinite macerals, it might, instead, be quite reactive.

The significant liquefaction yields that are presumed to have been derived from some macrinite-rich samples may attest to a contribution to conversion from this maceral. However, results from durains, and splint and cannel coals, which may contain large amounts of macrinite, generally have been variable (50). The conversion which we achieved with a coal containing 21% macrinite indicated that there was a contribution to the liquid products of batch hydrogenation from this maceral. Further, the residues examined from runs made with this coal at a series of temperatures contained no distinguishable macrinite product once a temperature of 425°C had been reached (66).

Because of the possibility that the response of macrinite is variable, then there may be justification for the recognition of a

Table IV. Liquefaction of Inertine-rich Samples

	V	E	F	SF	MA	MI	Ash, %	R_o max(V)	R_o max(F+SF)	Conversion‡, % daf
PSOC 303 Callide seam Callide, Qld.	15	2*	4	73	1	3	15.3	0.56	1.26	40
PSOC 304 Big seam Blair Athol, Qld.	31	2	14	44	5	4	6.9	0.64	1.49	41
PSOC 261, Fusain Illinois No. 6 Saline Co., Ill.	21	0	65	12	0	2	27.9	0.72	3.50	25
PSOC 261A, Fusain Illinois No. 6 Saline Co., Ill.	17	0	68	14	0	0	24.2	—	3.52	15
PSOC 262, Fusain Illinois No. 6 Williamson Co.,Ill.	16	0	69	14	0	0	20.6	0.66	4.18	21
PSOC 263, Fusain Illinois No. 6 Peoria Co., Ill.	8	0	48	44	0	0	2.1	0.47	3.05	13.5
PSOC 264, Fusain Colchester No. 2 Fulton Co., Ill.	10	0	62	27	0	0	22.6	—	4.23	12

* also contains 3% cutinite

V: vitrinite   E: exinite   F: fusinite   SF: semifusinite   MA: macrinite   MI: micrinite

‡ conversion to ethyl acetate solubles and gases at 400°C for one hour

semi-macrinite category in petrographic analysis, since the degree
of inertness could be related to reflectance, as we have shown is
the case with the series fusinite-semifusinite.

The reactive role of liptinite macerals in liquefaction has
been partially documented (50,68). However, recent work has shown
that unaltered sporinite often is encountered in the residues from
both batch and continuous liquefaction runs. For example, sporinite was a common component in the residues of a high volatile A
bituminous coal after hydrogen-transfer runs at 400° for 30 minutes (70). In spite of the relative unreactivity of the sporinite
in this instance, the vitrinite clearly had reacted extensively
because vitroplast was the predominant residue component. The
dissolution rate of sporinite from some coals, even at 400°C, may
be somewhat less than that of vitrinite.

In contrast to sporinite, resinite from a Utah high volatile
A bituminous coal reacted rapidly and more completely than the
corresponding vitrinite. Table V shows the conversion levels
achieved for a concentrate containing 75% resinite (mineral-free
basis) reacted under relatively mild conditions. The results are
curious. A fairly respectable level of conversion is achieved in
15 minutes at 350°C (under which conditions the associated vitrinite would presumably show little conversion), but longer times
and a temperature of 370° have little further effect; even raising
the temperature to 400° does not show a major increase in conversion.

Table V. Conversion of Utah Resinite Concentrate

Temperature (°C)	Time (min)	Conversion (% daf[*])
350	15	59
370	15	60
370	30	62
400	15	73

*For definition see Table IV

The residue produced from the 350°C run contained discernible
resinite particles. In contrast, examination of the fluorescence
of residues from the two 370° runs in blue light showed that little resinite was left undissolved other than that incorporated
within a matrix of other macerals. Instead, a diffuse fluorescence had been imparted to the epoxy resin embedding medium. Presumably, the epoxy was able to dissolve some of the liquefied
resin remaining after extraction with ethyl acetate. In the
residue from the run at 400°C, only one discrete resinite particle
was observed among the many coal particles embedded in the epoxy
polymer. It appears that in a short time at 350°, most, but not
all, of the resinite undergoes liquefaction. All other material
in the sample needs considerably more severe treatment.

2.6 Accumulation of Reactor Deposits

It is well known that during liquefaction there is always some amount of material which appears as insoluble, residual solids (65,71). These materials are composed of mixtures of coal-related minerals, unreacted (or partially reacted) macerals and a diverse range of solids that are formed during processing. Practical experience obtained in liquefaction pilot plant operations has frequently shown that these materials are not completely eluted out of reaction vessels. Thus, there is a net accumulation of solids within vessels and fluid transfer lines in the form of agglomerated masses and wall deposits. These materials are often referred to as reactor solids. It is important to understand the phenomena involved in reactor solids retention for several reasons. Firstly, they can be detrimental to the successful operation of a plant because extensive accumulation can lead to reduced conversion, enhanced abrasion rates, poor heat transfer and, in severe cases, reactor plugging. Secondly, some retention of minerals, especially pyrrhotites, may be desirable because of their potential catalytic activity.

In absolute terms, the quantities of reactor solids found in various processes do vary considerably. The rate of accumulation is related to several factors, such as coal characteristics, recycle solvent quality and reactor design. However, it can be stated in general terms that liquefaction of low rank coals (subbituminous C and lignites) does result in higher rates of accumulation of solids than do similar operations with bituminous coals. For example, during normal operations of the SRC-I pilot plant at Wilsonville, Ala., it has been found that the amount of solids retained varies from about 0.2-0.5 wt.% (moisture-free) for bituminous coals to 1.0-1.9 wt.% (moisture free) for a subbituminous C coal (Wyodak) (72). Exxon also reports much larger accumulations for lignites and subbituminous coals than those found for bituminous coals (73).

Perhaps the most important components of reactor solids are those that are generated during processing rather than those that are derived from inert minerals (quartz, clays) and macerals (fusinites, etc.) in the feed coal (74). The retention of these 'formed' materials is more difficult to predict from the characteristics of the feed and, hence, control in liquefaction processes. In most cases, the inert materials are merely held together by matrices of the internally generated solids. It is important to recognize that there are two very distinct types of material generated by coal liquefaction; namely carbonaceous (coke-like) solids and carbonates.

Carbonaceous solids appear as a result of retrogressive reactions, in which organic thermal fragments recombine to produce insoluble semi-cokes (59,65). Coke formation is observed during liquefaction of all coals and its extent can vary widely according to the coal, the reaction solvent, and reaction conditions. The predominant inorganic species produced during the process of coal

liquefaction are calcium carbonates. This phenomenon only occurs
during processing of low rank (subbituminous C and lignitic) coals.
    In the following, we shall discuss reactor solids in terms of
observed behavior of bituminous coals *versus* those of lower rank.
Examination of reactor solids obtained from bituminous coals (65,
74,75) nearly always reveals that they are predominantly composed
of mineral species (Table VI) that are derived from the minerals
occurring in the coals. These solids usually yield more than 60%
ash; the only exceptions are found when severe coking has occurred (74). The most common minerals identified in reactor solids
from bituminous coals are quartz, pyrrhotite and anhydrite. Minor
constituents differ among solids from different coals, but generally consist of clays, calcium-containing minerals (carbonates and
sulfates), iron-containing minerals (sulfides and carbonate), and
titanium oxides (65,74). We have shown that minerals such as
quartz, calcite and rutile are essentially unchanged by liquefaction (74). Clays and calcium sulfates appear to undergo dehydration and fragmentation. There is some doubt concerning the occurrence of calcium sulfates because it has been noted that they
appear in greater concentrations in reactor solids than in the
corresponding feed coals. The possibility of sulfate formation
cannot be ruled out, though a mechanism of formation is hard to
envisage. For all bituminous coal reactor solids studied, the
most striking mineralogical change is the reduction of pyrite to
pyrrhotite.
    For these types of reactor solids, the carbonaceous solids
content varies usually from about 20 to 40%. The components of
these solids are listed in Table VII. Optical examination of the
solids has shown that they are primarily composed of mixtures of
semi-cokes formed during liquefaction by retrogressive reactions
with chars derived from macerals. Unreacted macerals comprise
only a small fraction of these solids (65,74,75).
    Extensive studies have been made into the propensity of
various SRC's and SRC fractions to undergo retrogressive reactions (65,75). In these experiments, a selection of SRC's and SRC
components was heated to the desired temperature at elevated pressures (5000 psi). Coking propensity was defined as the amount of
pyridine insolubles produced under the selected conditions. Results from these studies indicated that the tendency of SRC to
coke is dependent on the characteristics of the feed coal. Under
comparable pyrolysis conditions, a whole SRC obtained from a Wyodak
(subbituminous C) coal does undergo retrogressive reactions slightly more readily than those derived from high volatile bituminous
coals. It has been suggested that this is a result of higher
concentrations of oxygen functionality in the Wyodak SRC. However,
it should be noted that coke formation is often dominated by factors that may mask the role of coal characteristics, e.g. availability of hydrogen donors, temperature, pressure and degree of
agitation. One important observation made on Wyodak reactor
solids is that the carbonaceous materials do tend to contain a
relatively large proportion of mesophase-derived semi-cokes (74).

Table VI. Mineralogical Compositions of Reactor Solids
           Determined by X-ray Diffraction (74)

Coal	Major Constituents	Minor Constituents	Ash from Reactor Solids wt% Dry Basis
W. Ky. Nos. 9/14 (Bituminous)	quartz, pyrrhotite, anhydrite	calcite, gypsum, illite, kaolinite, montmorillonite, rutile	61-71
Ill. No. 6 (Bituminous)	quartz, pyrrhotite, anhydrite	aragonite, bassanite, calcite, gypsum, illite, kaolinite, montmorillonite, rutile, siderite	71-80
Wyodak (sub-bituminous)	calcite, vaterite, quartz	anhydrite, kaolinite, pyrrhotite	60-70

Table VII. Carbonaceous Components of Reactor Solids (74)

1. Materials formed during liquefaction

    Pitch-like solids and isotropic semi-cokes
    Anisotropic semi-cokes

2. Unreacted macerals

    Fusinite
    Semifusinite
    Vitrinite
    Mixed-maceral particles

The tendency to coke formation also increases again, as might be expected, in the liquefaction of medium and low volatile bituminous coals.

It will be seen that, unfortunately, not very much can be said at the present time about the role of coal characteristics in determining agglomeration behavior in liquefaction. Effects do vary widely in different experiments, particularly in continuous flow pilot plants, but it is often difficult to disentangle effects of coal characteristics from effects due to differences or changes in reaction conditions. In a pilot plant, it is almost impossible to maintain precisely constant conditions for a length of time, and perturbations may trigger the onset of more severe coking. It appears that some coals are more susceptible than others to such upsets in conditions, but the causes in coal composition and geochemistry have not yet been identified. Much of the laboratory work in this field has related to the behavior of SRC samples previously prepared in a continuous flow reactor, rather than to the coal itself in a liquefaction reactor.

The most important feature of reactor deposits obtained from low rank coals is the formation of calcium carbonate (Table VI) as calcite and/or metastable vaterite (65,74,75,76,77). Optical and SEM studies on these solids indicates that carbonate is precipitated and is not related to calcite grains which may occur in the feed coal (65,73,74,77). It has been shown that during liquefaction of subbituminous and lignitic coals, calcium carboxylates in the coal structure decompose to produce calcium carbonate. In these coals, almost the entire calcium content is present in an ion-exchange form (as calcium carboxylates). Determination of the amounts of ion-exchangeable calcium by exchange with other cations, e.g., barium, indicates their extent in several low rank coals, as seen in Table VIII (65,75). In all cases, the amount of exchangeable calcium represents a large proportion (>80%) of the total calcium content. Additionally, the calcium content of these coals is much larger than those generally found for higher rank coals (65, 75). Of course, the quantity of ion-exchangeable calcium found in bituminous coals is extremely small because these coals contain little or no carboxylic acid functionality. Thus, the problem of calcium carbonate deposits is limited to liquefaction of low rank coals. A semi-quantitative relationship can be found between the amount of reactor solids formed and the ion-exchangeable calcium content of feed coals (65). Exxon workers have shown a linear relationship between the total calcium content of a series of coals and the quantity of calcium carbonate deposited in a pilot plant reactor (73). From these data, it is evident, again, that lignites and subbituminous coals yield by far the greatest amounts of carbonate deposits. They have also found that a Texas lignite can produce deposits of a sodium-magnesium carbonate (eitelite). Presumably, this carbonate deposit is related to exchangeable sodium and magnesium cations in that lignite.

Table VIII. Calcium Distribution in Selected Coals

Source of Coal	Total Ca^{++} wt% (Dry Coal)	Exchangeable Ca^{++} wt% (Dry Coal)
N. Dakota (lignite)	1.84	1.76
Wyoming (subbit. C)	1.20	1.17
Wyoming (subbit. C)	1.06	0.99
Wyoming (subbit. C)	1.03	0.86
Illinois No. 6 (bituminous)	0.30	0.17

A recent study in these laboratories (75) on calcium carbonate precipitation from Wyodak coal has confirmed the relationship between ion-exchangeable calcium and the appearance of calcium carbonates during liquefaction. These experiments were performed on samples of the subbituminous coal which had been demineralized, to ensure that all carboxylic acid groups were in the acidic form, and subsequently exchanged with varying amounts of calcium ions.

In addition to the content of ion-exchangeable calcium, other factors must be considered when the rate of accumulation is in question. In order that the precipitates be retained in reaction vessels, it is necessary that they grow to a sufficient size to preclude elution. This condition is achieved in reactor configurations where residence time is relatively long. Alternatively, if turbulent conditions prevail, as in the H-Coal reactor, the precipitates may be abraded or not allowed to grow, so that retention would be inhibited, though their formation will not be prevented.

Literature Cited

1. Spackman, W.; Barghoorn, E. S. In "Coal Science", Amer. Chem. Soc. Adv. Chem. Ser. 55, Peter H. Given, Ed., Washington, D. C., 1966, pp. 695-707.

2. Phillips, T. L.; Mickish, D. J.; Kunz, A. B. In "Interdisciplinary Studies of Peat and Coal Origins", Geol. Soc. Amer. Microform Publication 7, P. H. Given and A. D. Cohen, Eds., 1977, pp. 18-49.

3. Given, P. H. In "Advances in Organic Geochemistry 1971", H. Wehner and H. von Gaertner, Eds., Vieweg und Sohn, Branschweig, Germany, 1972, p. 69.

4. Given, P. H.; Dickinson, C. H. In "Soil Biochemistry", E. A. Paul and A. D. McLaren, Eds., Marcel Dekker, Inc., New York, 1975, v. 3, pp. 123–212.

5. Mackowsky, M.-Th. Chapter 14 in "Coal and Coal-Bearing Strata", D. Murchison and T. S. Westoll, Eds., Oliver & Boyd, Edinburgh, 1968, p. 325.

6. Brooks, J.; Shaw, G. Nature, 1968, 219, 532–533.

7. Brooks, J.; Shaw, G., Grana, 1978, 17, 91–97.

8. Hopkinson, Shirley M. Quart. Rev. Chem. Soc., 1969, 23, 98.

9. Hillis, W. E., Ed. "Wood Extractives and their Significance to the Pulp and Paper Industry", Academic Press, New York, London, 1962, 513 pp.

10. Teichmüller, M. In "Coal Petrology", by E. Stach, G. H. Taylor, M.-Th. Mackowsky, D. Chandra, M. Teichmüller and R. Teichmüller, Gebrüder Borntraeger, Berlin, 1975, pp. 194–198.

11. Chen, Angeli S-H. "Flavonoid Pigments in the Red Mangrove, *Rhizophora Mangle L.*, of the Florida Everglades and in the Peat Derived from it", M.S. Thesis, Pennsylvania State University, 1971, 233 pp.

12. Harborne, J. B. "Comparative Biochemistry of the Flavonoids", Academic Press, London and New York, 1967.

13. Bate-Smith, E. C. J. Linnean Soc. London, Botany, 1962, 58, 95.

14. Harborne, J. B., Ed. "Biochemistry of Phenolic Compounds", Academic Press, London, 1964.

15. Teichmüller, M.; Teichmüller, R. In "Coal Science", Amer. Chem. Soc. Adv. Chem. Ser. 55, Peter H. Given, Ed., Washington, D. C., 1966, pp. 133–155.

16. Teichmüller, M.; Teichmüller, R. Chapter 11 in "Coal and Coal-Bearing Strata", D. Murchison and T. S. Westoll, Eds., Oliver & Boyd, Edinburgh, 1968, pp. 233–267.

17. Karweil, J. In "Advances in Organic Geochemistry 1968", P. A. Schenk and I. Havenaar, Eds., Pergamon, 1969, p. 59.

18. van Krevelen, D. W. Chapter 14 in "Coal", Elsevier, Amsterdam, 1961, pp. 286–300.

19. Hill, G. R.; Hariri, H.; Reed, R. I.; Anderson, L. L. In
     "Coal Science", Amer. Chem. Soc. Adv. Chem. Ser. 55, Peter
     H. Given, Ed., Washington, D. C., 1966, p. 427.

20. Szladow, A. J. "Some Aspects of the Mechanism and Kinetics
     of Coal Liquefaction", Ph.D. Thesis, Pennsylvania State Uni-
     versity, 1979, 172 pp.

21. Szladow, A. J.; Given, P. H. Ind. Eng. Chem. (Proc. Des.
     Dev.), paper submitted August 1979.

22. Tissot, B.; Welte, D. "Petroleum Formation and Occurrence",
     Springer-Verlag, Berlin, 1978, pp. 123-147.

23. Mott, R. A. Contribution to Discussion, "Proc. Conf. on
     Ultra-fine Structure of Coals and Cokes", British Coal Utili-
     zation Research Association, 1944, pp. 156-159.

24. White, E.; Towers, G. H. N. Phytochemistry, 1967, 6, 663.

25. Murchison, D. G. In "Coal Science", Amer. Chem. Soc. Adv.
     Chem. Ser. 55, Peter H. Given, Ed., Washington, D. C., 1966,
     p. 307.

26. Teichmüller, M. In "Coal Petrology", by E. Stach, G. H.
     Taylor, M.-Th. Mackowsky, D. Chandra, M. Teichmüller and R.
     Teichmüller, Gebrüder Borntraeger, Berlin, 1975, pp. 218-220.

27. Austen, D. E. G.; Ingram, D. J. E.; Given, P. H.; Binder, C.
     R.; Hill, L. W. In "Coal Science", Amer. Chem. Soc. Adv.
     Chem. Ser. 55, Peter H. Given, Ed., Washington, D. C., 1966,
     pp. 344-362.

28. Given, P. H.; Binder, C. R. In "Advances in Organic Geochem-
     istry 1964", G. D. Hobson and M. C. Louis, Eds., Pergamon,
     1966, p. 147.

29. Given, P. H.; Peover, M. E.; Wyss, W. F. Fuel, 1965, 44,
     425.

30. Bent, R.; Brown, J. K. Fuel, 1961, 40, 47.

31. Hirsch, P. B. Phil. Trans. Roy Soc., 1960, 252A, 68.

32. Binder, C. R. "Electron Spin Resonance: Its Application to
     the Study of Thermal and Natural Histories of Organic Sedi-
     ments", Ph.D. Thesis, Pennsylvania State University, 1965,
     129 pp.

33. Gray, D.; Barrass, G.; Jezko, J. This volume, p. 35.

34. Durie, R. A. This volume, p.53.

35. Taylor, G. H.; Mackowsky, M.-Th.; Alpern, B. Fuel, 1967, 46, 431.

36. Neavel, R. C. Amer. Chem. Soc. Fuel Chem. Div. Preprints 24(1), pp. 73-82 (papers presented at Honolulu, Hawaii, April 2-6, 1979).

37. Francis, W. "Coal", Arnold, London, 1961, pp. 300, 302, 432.

38. Given, P. H. In "Environmental Organic Chemistry", Specialist Periodical Rept., G. Eglinton, Senior Reporter, Chem. Soc., London, 1975, p.

39. Teichmüller, M. Fortschr. Geol. Rheinl. Westfälen, 1974, 24, pp. 65-112.

40. Given, P. H.; Miller, Robert N. Abstracts with Programs, Geol. Soc. Amer., 1971 Annual Meeting, 3(7), p. 580. [Typescript of full text available on request].

41. Brock, Thomas D. "Biology of Microorganisms", Prentice-Hall, Englewood Cliffs, N. J., 1979, 3rd Edition, pp. 434-439.

42. Kaplan, I. R.; Emery, K. O.; Rittenburg, S. C. Geochim. et Cosmochim. Acta, 1963, 27, 297-331.

43. Spackman, W.; Cohen, A. D.; Given, P. H.; Casagrande, D. J. "The Comparative Study of the Okefenokee Swamp and the Everglades-Mangrove Swamp-Marsh Complex of Southern Florida, Field guidebook printed for Geol. Soc. Amer. Pre-convention field trip, 15-17 November 1974 (subsequently published by Coal Research Section, Pennsylvania State University, 1976), 403 pp.

44. Damberger, H. H. Econ. Geol., 1971, 66, 488.

45. Waddell, C.; Davis, A.; Spackman, W.; Griffiths, J. C. "Study of the Interrelationships among Chemical and Petrographic Variables of United States Coals", Tech. Rept. 9 to U.S. Department of Energy from Coal Research Section, Pennsylvania State University, Rept. FE-2030-TR 9, 1978.

46. Nichols, G. D. Chapter 12 in "Coal and Coal-Bearing Strata", D. Murchison and T. S. Westoll, Eds., Oliver & Boyd, Edinburgh, 1968, p. 269.

47. Miller, R. N. "A Geochemical Study of the Inorganic Constituents in Some Low-Rank Coals", Ph.D. Thesis, Pennsylvania State University, 1977, 314 pp.

48. Ignasiak, B.; Carson, D.; Szladow, A.; Berkowitz, N. This volume, p. 97.

49. Fisher, C. H.; Eisner, A. Ind. Eng. Chem., 1937, 29, 939-945.

50. Fisher, C. H.; Sprunk, G. C.; Eisner, A.; O'Donnell, H. J.; Clarke, L.; Storch, H. H. "Hydrogenation and Liquefaction of Coal. II. Effect of Petrographic Composition and Rank of Coal", U.S. Bureau of Mines Tech. Paper 642, 1942, 162 pp.

51. Glenn, R. A.; Basu, A. N.; Wolforth, J. S.; Katz, M. Fuel, 1950, 29, 149-159.

52. Hirst, L. L.; Storch, H. H.; Fisher, C. H.; Sprunk, G. C. Ind. Eng. Chem., 1940, 32, 864-871.

53. Hirst, L. L.; Storch, H. H.; Fisher, C. H.; Sprunk, G. C. Ind. Eng. Chem., 1940, 32, 1372-1379.

54. Wu, W. R. K.; Storch, H. H. "Hydrogenation of Coal and Tar", U. S. Department of the Interior, Bureau of Mines Bull. 633, 1968, 195 pp.

55. Ergun, S. In "Coal Conversion Technology", C. Y. Wen and E. Stanley Lee, Eds., Addison-Wesley Publishing Co., Reading, Mass., 1979, pp. 1-56.

56. Wright, C. C.; Sprunk, G. C. Penn. State Coll. Min. Ind. Exp. Stat. Bull. 26, 1939, 32 pp.

57. Given, P. H.; Cronauer, D. C.; Spackman, W.; Lovell, H. L.; Davis, A.; Biswas, B. Fuel, 1975, 54, 34-39.

58. Whitehurst, D. D.; Farcasiu, M.; Mitchell, T. O.; Dickert, J. J. "The Nature and Origins of Asphaltenes in Processed Coals", Annual Rept. AF-480 from Mobil Res. & Dev. Corpn. to Electric Power Research Institute, 1977.

59. Whitehurst, D. D.; Farcasiu, M.; Mitchell, T. O.; Dickert, J. J. "The Nature and Origins of Asphaltenes in Processed Coals: Chemistry and Mechanisms of Coal Conversion to Clean Fuel", Annual Rept. for 1978 from Mobil Res. & Dev. Corpn. to Electric Power Research Institute, AF-1298, Vol. 2.

60. Abdel-Baset, M. B.; Yarzab, R. F.; Given, P. H. Fuel, 1978, 57, 89-94.

61. Yarzab, R. F.; Given, P. H.; Davis, A.; Spackman, W. Fuel, 1980, 59, 81-92.

62. Yarzab, R. F.; Baset, Z.; Given, P. H. Geochim. et Cosmochim.
    Acta, 1979, 43, 281–287.

63. Given, P. H.; Schleppy, R.; Sood, Ajay. Fuel, 1980 (in the
    press).

64. Granoff, B.; Baca, T. M.; Thomas, M. G.; Noles, G. P. "Chemi-
    cal Studies on Synthoil: Mineral Matter Effects", Sandia Labs.
    Energy Rept. No. SAN-78-1113, 1978.

65. Walker, P. L.; Spackman, W.; Given, P. H.; Davis, A.; Jenkins,
    R. G.; Painter, P. C. "Characterization of Mineral Matter in
    Coals and Coal Liquefaction Residues", Annual Rept. AF-832
    from Pennsylvania State University to Electric Power Research
    Institute, 1978.

66. Mitchell, G. D.; Davis, A.; Spackman, W. In "Liquid Fuels
    from Coal", Rex T. Ellington, Ed., Academic Press, 1977, pp.
    255–270.

67. Neavel, R. C. Fuel, 1976, 55, 237.

68. Davis, A.; Spackman, W.; Given, P. H. Energy Sources, 1976,
    3(1), 55.

69. Shapiro, N.; Gray, R. J.; Eusner, G. R. A.I.M.E. Proc. Blast
    Furnace, Coke Oven & Raw Materials, 1961, v. 20, pp. 89–112.

70. Given, P. H.; Spackman, W.; Davis, A.; Walker, P. L.; Lovell,
    H. L.; Coleman, M.; Painter, P. C. "The Relation of Coal Char-
    acteristics to Liquefaction Behavior", Quart. Tech. Prog. Repts.
    for period Jan.-June 1978 to U.S. Department of Energy under
    Contract No. EX-76-C-01-2494, Rept. Nos. FE-2494-7/8, 1978.

71. Walker, P. L.; Spackman, W.; Given, P. H.; White, E. W.;
    Davis, A.; Jenkins, R. G. "Characterization of Mineral Matter
    in Coals and Coal Liquefaction Residues", Rept. No. EPRI AF-417,
    RP366-1 from Pennsylvania State University to Electric Power
    Research Institute, June 1977.

72. Lewis, M. E.; Weber, W. H.; Usnick, G. B.; Hollenack, W. R.;
    Hooks, H. W. Annual Rept. for 1976 for Operation of Wilson-
    ville SRC Pilot Plant from Catalytic, Inc. to Electric Power
    Research Institute and U.S. Department of Energy, Project
    1234-1-2, November 1977.

73. Epperly, W. R. (Project Director). "EDS Coal Liquefaction
    Process Development—Phase IV", Annual Tech. Rept. for July
    1978-June 1979 from Exxon Research and Engineering Company
    to U.S. Department of Energy and Electric Power Research In-
    stitute under Contract No. EF-77-A-01-2893, September 1979.

74. Wakeley, L. D.; Davis, A.; Jenkins, R. G.; Mitchell, G. D.;
     Walker, P. L. Fuel, 1979, 58, 379-385.

75. Walker, P. L.; Spackman, W.; Given, P. H.; White, E. W.;
     Davis, A.; Jenkins, R. G.; Painter, P. C. "Characterization
     of Mineral Matter in Coals and Coal Liquefaction Residues",
     Final Rept. from Pennsylvania State University to Electric
     Power Research Institute, 1980 (in preparation).

76. Brunson, R. J.; Chaback, J. J. Chem. Geol., 1979, 25(4),
     333-338.

77. Stone, J. B.; Trachte, K. L.; Poddar, S. K. Amer. Chem. Soc.
     Fuel Chem. Div. Preprints 24(2), pp. 255-262 (papers presented
     at Honolulu, Hawaii, April 2-6, 1979).

RECEIVED May 16, 1980.

South African Coals and Their Behavior During Liquefaction

D. GRAY, G. BARRASS, J. JEZKO, and J. R. KERSHAW

Fuel Research Institute of South Africa, P.O. Box 217, Pretoria, South Africa 0001

South African coals differ from most Northern Hemisphere coals in their geological age, unusual petrology and their high mineral matter content. If these coals are to be used for conversion to synthetic fuels then criteria must be found to enable predictions of their behaviour under liquefaction conditions to be determined. This paper describes the hydrogenation of a number of South African coals using two different techniques, to ascertain whether well known coal properties can be used to predict their hydrogenation behaviour.

The effect of the mineral matter content and of the inorganic sulphur content on the hydrogenation of coal were also studied.

GEOLOGICAL ORIGIN OF SOUTH AFRICAN COALS

The coal deposits of South Africa were formed during the Permian age just after a retreat of glaciation. This makes it almost certain that the climate was temperate rather than subtropical and may explain the differences between the plant life in South Africa at that time and the flora of the North American Carboniferous era. The predominance of the inertinite maceral group in South African coals is indicative of drier swamp conditions in which rotting processes as well as peatification played a more dominant role than in the formation of the humic coals of Europe (1). South African coals were thought to be deposited in deltaic or fluvial environments where fluctuations in water level may have caused deposition of large quantities of mineral matter. South African coals generally have not reached a very high rank although in Natal anthracites are found. This rank increase seems to be due to metamorphism brought about by dolerite intrusions rather than by stratigraphic depth.

GEOGRAPHICAL LOCATION AND COAL RESERVES OF SOUTH AFRICA

A map showing the coal fields of the Republic of South Africa is shown in Figure 1. The main coalfields lie in the Highveld area of the Orange Free State, South-Eastern Transvaal

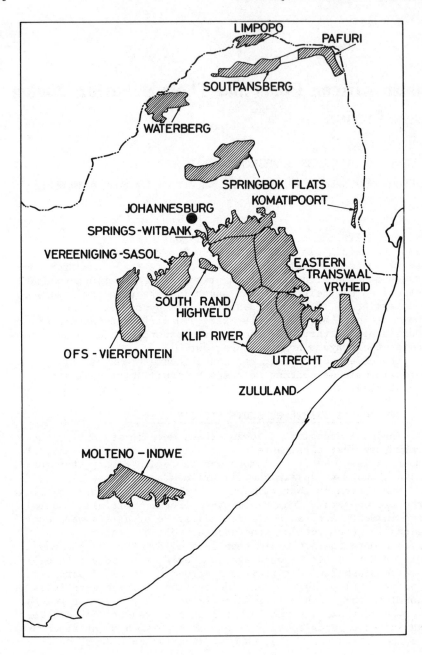

Figure 1.   Coalfields of South Africa

and Natal. Other coalfields are located in the Northern part of
the country, the Waterberg field on the Botswana border and the
Limpopo and Pafuri fields on the borders of Zimbabwe/Rhodesia
and Mozambique. The largest field is the Highveld field (25,407
million metric tons of raw bituminous mineable coal in situ)
followed by the Witbank and Waterberg fields (approximately
17,000 million metric tons) (2).

Table I   Raw Bituminous Coal Resources

Ash %	5-10	10-15	15-20	20-25	25-30	30-35	Total
Resources Million Metric Tons	20	1,499	8,620	15,220	18,565	37,350	81,274

Table I shows the raw bituminous coal resources figures in
millions of metric tons at various ash contents (2). The total
figure of 81,274 million metric tons is for the total resources
of raw bituminous coal mineable in situ down to 300 meters.
This total is the sum of the proven, indicated and inferred
reserves.
Mineable coal in situ is that portion of the coal in situ
which can be mined by existing techniques. The Petrick commission
report (2) arrives at a figure of 24,915 million metric tons of
raw bituminous coal extractable by underground mining. Extractable coal is defined as that portion of the mineable coal in
situ which is extractable in prevailing or slightly less rigorous
economic conditions. The figure of 24,915 million metric tons
for a raw bituminous coal has been substantially added to since
1975 by further exploration and could well be of the order of
28,000 million metric tons by this time. It must also be emphasized that reserves are dynamic, as higher prices or improved
technology may allow the exploitation of deposits which are
presently not regarded as reserves.
Table I shows that most of the South African bituminous coal
contain high quantities of mineral matter which is often intimately associated with the organic matter of the coal. About
half of the resources yield between 30 and 35 per cent ash.

LIQUEFACTION BEHAVIOUR OF A SELECTION OF SOUTH AFRICAN COALS

Experimental   Two different experimental procedures were used
in this study, to identify the coal properties of importance in
coal conversion which are independent of processing conditions.
These were:
(i)   'Dry' Hydrogenation using a semi-continuous 'hot-rod'
         reactor.
(ii)   Slurried Hydrogenation using a rotating autoclave.

(i)   'Hot-Rod' Method
        This method was similar to that used by Hiteshue et al (3).
In this method sand (50 g, mesh 0.42 - 0.15 mm) was mixed with
the coal (25 g, mesh 0.5 - 0.25 mm). The addition of sand to
the coal helped to prevent agglomeration (4). All the experiments used an aqueous solution of stannous chloride impregnated
on the coal as a catalyst. The amount of catalyst added on a
tin basis was 1% of the mass of the coal. These mixtures were
placed in a 'hot-rod' reactor and heated to $500^{\circ}C$ at a heating
rate of $200^{\circ}C$ per minute. Residence time at temperature was 15
minutes. Hydrogen at a flow rate of 22 liters/minute and a
pressure of 25 MPa was continously passed through the fixed bed
of coal/sand/catalyst. The volatile products were collected in
high-pressure cold traps. A schematic of the apparatus used is
shown in Figure 2.

(ii)   Rotating Autoclave Method
        The reactor was a 1 liter stainless steel rotating autoclave.
In these experiments the ratio of anthracene oil to coal was 3:1.
Coal (50 g) impregnated with catalyst (1% Sn as $SnCl_2$) was mixed
with sand (200 g). The autoclave was pressurized with hydrogen
to 10 MPa at room temperature and heated (ca $7^{\circ}C$/minute) to the
final reaction temperature ($450^{\circ}C$). The pressure at reaction
temperature was approximately 25 MPa.
        The product was washed from the cooled 'hot-rod' reactor
system or autoclave with toluene. The solid residue was extracted with boiling toluene in a soxhlet extractor for 12 hours.
        The overall conversion of coal to liquid and gaseous
products was obtained from the formula:

Percentage conversion = 100 $\left\{ 1 - \dfrac{\text{Organic material in the residue}}{\text{Organic material in the coal}} \right\}$

        In the case of the 'hot-rod' reactor experiments, the
toluene solutions were combined and the toluene removed under
reduced pressure. n-Hexane (250 ml) was added to the extract
and it was allowed to stand for 24 hours with occasional shaking.
The solution was filtered to leave a residue (asphaltene) and
the hexane was removed from the filtrate under reduced pressure
to give the oil.

Properties of the coals used

        The chemical and petrographic properties of the twenty coals
used in the hydrogenation experiments are shown in Tables II and
III. Mineral matter was determined directly using a radio
frequency low temperature plasma asher at medium power rating for
approximately 48 h per coal. The volatile matter was corrected
for the effects of the mineral matter by applying the equation
used by Given et al (5).
        The mean maximum reflectance of vitrinite ($\bar{R}o$ max) is the
mean of one hundred determinations.
        The percentage of reactive macerals on a volume basis was

Table II Proximate and Ultimate Analyses of Coals Used in the Liquefaction Experiments

Coal	Proximate Analyses (Air Dried Basis)				Ultimate Analyses (D.A.F. Basis)				H/C Atomic Ratio	MM[1]	VMc[2]
	% H$_2$O	% Ash	% Vol. Mat.	% Fix. Carb.	% C	% H	% N	% S			
Matla	6.0	10.9	35.3	47.8	78.8	5.5	2.0	0.5	0.83	10.3	40.1
Waterberg	3.4	12.7	34.8	49.1	80.6	5.4	1.5	1.0	0.81	15.2	40.2
New Wakefield	4.9	14.9	32.8	47.4	79.1	5.4	2.1	2.3	0.82	19.0	38.3
Kriel	3.8	20.2	30.3	45.7	79.1	5.3	2.0	2.9	0.80	22.3	35.9
Spitzkop	3.2	12.7	32.7	51.4	81.6	5.3	2.1	1.3	0.78	16.0	37.4
Landau	2.5	14.3	23.3	59.9	84.1	4.4	1.9	0.6	0.63	15.2	25.8
Koornfontein	2.6	20.6	27.9	48.9	82.9	5.5	2.0	0.8	0.79	22.8	32.9
Delmas	4.7	11.9	29.6	53.8	80.3	5.0	1.9	1.3	0.75	14.5	34.1
Sigma	7.1	38.8	19.8	34.3	74.3	4.5	1.7	0.9	0.73	45.0	28.0
Tvl. Nav.	2.6	14.3	26.3	56.8	83.0	5.0	2.1	0.8	0.72	17.2	29.9
Springbok	2.4	14.6	25.7	57.3	83.5	4.9	2.0	1.5	0.70	16.4	28.5
Vierfontein	7.2	27.6	20.8	44.4	76.5	4.7	2.0	1.8	0.69	31.9	26.4
Ballengeich	2.2	17.4	23.6	56.8	84.2	5.0	2.2	1.9	0.71	19.9	26.4
Cornelia	7.8	28.6	22.1	41.5	78.4	4.5	2.0	1.6	0.68	32.8	29.0
Phoenix	3.6	13.0	24.9	58.5	88.0	4.7	2.1	0.7	0.64	15.5	28.2
Wolvekrans	2.4	20.8	20.5	56.3	83.5	4.6	1.9	0.8	0.66	21.4	22.4
Newcastle	1.4	24.4	17.8	56.4	87.0	4.5	2.3	3.0	0.62	28.7	19.1
Natal Amm.	1.8	8.2	8.9	81.1	90.8	3.8	2.4	0.7	0.50	9.0	8.5
Utrecht	1.8	12.2	8.4	77.6	89.6	3.5	2.4	1.7	0.47	14.8	7.5
Balgray	2.0	10.0	6.1	81.9	90.9	3.0	2.2	1.2	0.40	11.3	5.1

1. Mineral Matter
2. Volatile Matter corrected to dry mineral matter free basis

Table III   Petrographic Composition of the Coals Used
                 (Mineral Matter Free Basis)

Coal	Fuel Research Institute V+E Volume %	$\bar{R}_o$ [1] max	South African Iron and Steel Corporation V+E[2] Volume %	RSF[3]	Total
Matla	84	0.656	84	4	88
Waterberg	94	0.720	94	1	95
New Wakefield	85	0.677	85	5	90
Kriel	74	0.685	79	9	88
Spitzkop	63	–	62	22	84
Landau	70	0.746	69	16	85
Koornfontein	61	0.787	61	22	83
Delmas	54	0.661	54	27	81
Sigma	32	0.599	32	53	85
Tvl. Nav.	42	0.748	42	33	75
Springbok	38	0.815	38	34	72
Vierfontein	35	0.619	35	33	68
Ballengeich	47	0.858	47	31	78
Cornelia	29	0.569	24	55	79
Phoenix	42	0.793	42	33	75
Wolvekrans	34	0.760	34	36	70
Newcastle	50	1.116	50	13	63

1   Mean Maximum Reflectance of Vitrinite
2   Vitrinite + Exinite
3   Reactive semi-fusinite

determined by two techniques. The vitrinite and exinite values
were obtained by the Fuel Research Institute and the total percentage of 'reactive' macerals was determined by the South
African Iron and Steel Corporation (ISCOR) (6). A number of
these coals contain large quantities of reactive semi-fusinite
(see Table III). This has important implications for the pre-
diction of technological behaviour and will be discussed later.

RELATIONSHIPS BETWEEN LIQUEFACTION BEHAVIOUR AND THE COAL COMPOSITION

The effect of the following coal property parameters was
studied in relation to liquid yields and conversions during
coal hydrogenation using both experimental procedures.
1. Volatile Matter Yield
2. H/C Atomic Ratio
3. 'Reactive' Maceral Content
4. Rank
5. Mineral Matter Content and Composition

Organic Coal Properties

Good correlations are obtained by plotting the percentage
total conversion of the coal expressed on a dry mineral matter
free basis (dmmf) against the corrected volatile matter content
for both the 'hot-rod' and autoclave modes (see Figures 3 and 4).
The slopes of the regression lines are very similar in both
modes and the square of the correlation coefficients are in both
cases very close to 1. The fact that the slopes are similar for
the results from the different experimental techniques, indicates
that thermal fragmentation is the overriding step in conversion
of the coal and that any differences are related to the mode of
stabilization of the reactive fragments formed.
    Figures 5 and 6 show that there is a good correlation between
the percentage conversion and the H/C atomic ratio of the coals
for both experimental procedures.
    Thus the corrected volatile matter yield and the atomic H/C
ratio both appear to be good parameters for assessing the
reactivity of the coals studied.
    There is also intercorrelation between the volatile matter
and the H/C atomic ratio for the South African coals studied.
Thus a good correlation between conversion yield and one of
these properties obviously implies a similar correlation with the
other property. The correlations between the volatile matter
yield and the 'reactive' maceral content and between the H/C
atomic ratio and the 'reactive' maceral content are not statis-
tically significant.
    No data on liquid yields are available for the autoclave
experiments because it is not possible to separate the product
oil, which results from coal liquefaction, from the anthracene
oil and its decomposition products. In the case of the 'hot-
rod' experiments this complication does not exist.

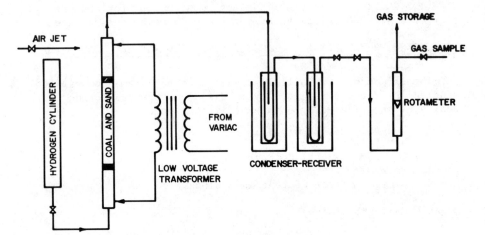

Figure 2.   Hot rod reactor

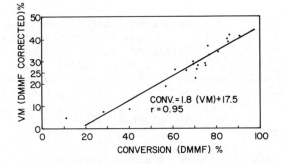

Figure 3.   Percentage conversion against volatile matter yield (hot rod mode)

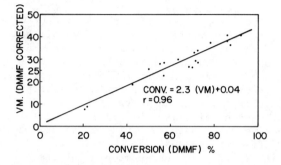

Figure 4. Percentage conversion against volatile matter yield (rotating autoclave mode)

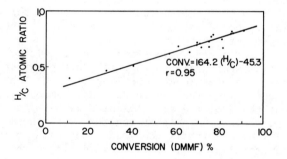

Figure 5. Percentage conversion against H/C atomic ratio (hot rod mode)

For the latter technique the yield of toluene solubles (oil plus asphaltene) is the most reliable of the liquid yield figures because of the uncertainty in estimating the asphaltene yield by precipitation with n-hexane (7). Figure 7 shows the variation in toluene soluble yield with H/C atomic ratio and with volatile matter yield of the coals. The best fit of the data for the toluene solubles against H/C atomic ratio is a power curve passing through the origin, whereas for the volatile matter yield a linear correlation was marginally better.

Figure 8 shows the effect of rank, as measured by the mean maximum reflectance of vitrinite, on the overall conversion. The conversion was highest for coals in a narrow Ro max range of between 0.65 - 0.70. Cudmore's data on Australian coals also appears to exhibit a maximum when reflectance data is plotted against conversion (8). It is difficult to interpret this data because of the large variation in the vitrinite and reactive semi-fusinite content of these coals. The reactivity of vitrinite and reactive semi-fusinite would be expected to vary with rank but to different degrees. For several of the lower rank coals vitrinite is only a minor component of the coal.

In Figure 9 the vitrinite + exinite content and the 'reactive' maceral content, as determined by ISCOR, are plotted against the total conversion for the 'hot-rod' technique. Figure 10 plots the same information for the autoclave results.

The different slopes for the lines of best fit for 'total reactives' and vitrinite + exinite reflects the special petrology of the majority of South African coals used in this study (see Table III). For these coals the 'reactives' contain a high proportion of semi-fusinite in the inertinite.

Several investigators have tried to characterize the behaviour of reactive semi-fusinite in coals during carbonization. Recently there have been reports dealing with the behaviour of this maceral under liquefaction conditions. Ammosov et al (9) classified one third of the semi-fusinite as being reactive during coking, whilst the balance together with the micrinite are inert. Taylor originally concluded that semi-fusinite and micrinite in Australian coals were inert during carbonization but later observed partial fusing of a transitional material between vitrinite and semi-fusinite (10, 11).

For American and European coking coals the behaviour of semi-fusinite is generally less important since only small quantities of this maceral are usually present. However, South African coal used in coke oven-blends contains as little as 40 per cent vitrinite and as much as 45 per cent reactive semi-fusinite (12). The partial reactivity of the semi-fusinite fraction during liquefaction of Australian coals has been reported by Guyot et al (13). They found that the low reflecting inertinite in two coals up to V_{14} (a reflectance from 1.40 to 1.49) was reactive. This agrees with the results of Smith and Steyn (12) who consider that the semi-fusinite fraction in South African coals up to V_{15} (1.50 - 1.59) can be reactive to coking.

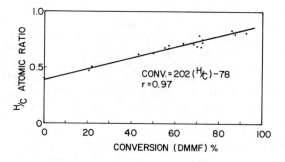

Figure 6. Percentage conversion against H/C atomic ratio (rotating autoclave mode)

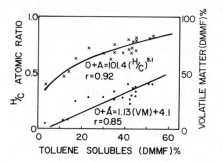

Figure 7. Percentage toluene solubles against H/C atomic ratio (×) and volatile matter yield (●)

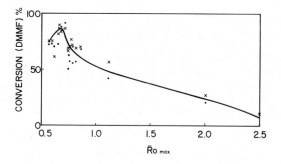

Fuel

Figure 8. Percentage conversion vs. rank (17): hot rod results (×); anthracene oil (●).

In addition, Shibaoka et al (14) during hydrogenation of a New South Wales coal reported that the inertinite with relatively low reflectance became partially liquefied.

On the basis of numerous petrographic analyses of South African coking coals varying in vitrinite content from 40 per cent up to 90 per cent Smith and Steyn (12) concluded that "semi-fusinite cannot be added to the reactives on a fixed arbitrary basis but need to be counted as a distinct group of reactives which can form up to 60 per cent of the total of semi-fusinite plus micrinite". Since the low reflecting unstructured, or only slightly structured, semi-fusinite in a coal has a significant role in the coking process, it is reasonable to assume that this maceral has an equally important role in lique-faction processes. For example a coal like Sigma having a con-ventionally assessed vitrinite + exinite content of only 32 per cent still gives a conversion yield, on hydrogenation, of 75 per cent (dmmf coal basis). The total 'reactive' maceral content of this coal is 85%.

The slopes of the regression lines for conversion yield against 'reactive' macerals for the 'hot-rod' and for the rota-ting autoclave modes of hydrogenation are shown by statistical analysis to be similar (compare Figures 9 and 10). This suggests that the relationship between total 'reactive' macerals and coal reactivity as measured by conversion is not dependent on the conversion technique.

However, coal reactivity as measured by total conversion to liquids and gases becomes less dependent on coal parameters as processing severity increases. The effect of process temperature in the 'hot-rod' reactor was studied using three coals of varying properties. These were Waterberg, Sigma and Landau. At 650°C the conversion yields of these coals were 89, 90 and 88 per cent of the coal (dmmf) respectively. Within experimental error the conversion yields had converged to the same value, whereas at 500°C the conversion yields were 85, 75 and 65 per cent respec-tively.

THE EFFECTS OF THE INORGANIC CONSTITUENTS

The effects of the inorganic constituents in the coal were studied in two ways. Firstly, to obtain samples with varying mineral matter content, a coal was subjected to a float and sink separation. These fractions were subsequently hydrogenated. The analyses of the float and sink fractions are shown in Table IV.

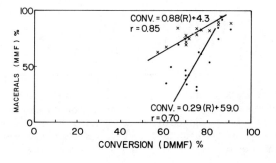

Figure 9. *Percentage conversion against vitrinite + exinite (●) and total reactive macerals (×) (hot rod mode) (17)*

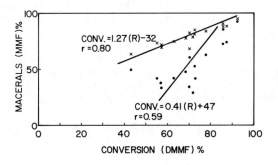

Figure 10. *Percentage conversion against vitrinite + exinite (●) and total reactive macerals (×) (rotating autoclave mode)*

Table IV Petrographic Analyses and Ash Yield of
          Float/Sink Fractions

Relative Density	Vitrinite	Exinite	Inertinite	Visible Minerals	Ash wt % air dried basis
			Volume %		
1.4 float	80.6	7.5	6.5	5.4	6.7
1.4 - 1.5	78.3	8.3	9.1	4.3	7.0
1.5 - 1.65	65.6	9.1	15.9	9.7	16.0
1.65 sink	66.6			33.4	45.3

It has been demonstrated that certain coal minerals, particularly iron compounds, catalyze the hydrogenation of coal-derived solvents (15). Mukherjee et al (16) hydrogenated float/sink fractions of an Indian coal and found that the conversion increased with the amount of mineral matter present in the fraction.

It is difficult, however, to assess precisely the effect that the mineral matter present in a coal has on its liquefaction behaviour. In float/sink fractions of the same coal, the petrographic constituents of the coal fractions usually change significantly, with more inertinite being found in the higher density fractions. Also the mineral matter composition and concentration changes from fraction to fraction, and there may be considerable variation in the mineral matter surface areas available for possible catalysis. In addition, the combination of the effects of increased mineral matter and decreased 'reactive' maceral content in the higher density fractions reduces the agglomeration tendency of the coal. This allows more effective diffusion in the system (4, 14).

The results obtained from the float/sink fractions are shown in Figure 11. It could be that the increase in oil yield obtained with the higher mineral matter fractions is due to the increase in sulfur content that varies from 0.5 per cent in the 1.4 float to 9 per cent in the 1.65 sink fraction. The significance of these results, at least as far as South African coals are concerned, is that a high mineral matter content does not necessarily mean poor performance during coal liquefaction. Indeed, this evidence suggests that the mineral matter can be beneficial in increasing both the conversion and liquid product yields. From a processing viewpoint, high mineral matter can create other problems, and a trade off between possible catalytic benefits and engineering process difficulties is necessary.

The second procedure studied the effects of the sulfur content of the coals during hydrogenation. A suite of unwashed coals was selected so that the only parameter to show significant variation was the total sulfur content. Of this total sulfur, approximately 1 per cent was organic, and the rest was inorganic sulfide. The maceral content and total mineral matter content of all these samples were very similar. Relevant analyses of this suite of coals is shown in Table V.

Table V   Analyses of Unwashed Coals Used to Determine the Effect of Pyrites

Coal	V+E %	Total Ash %	Moisture %	Sulfur %	VM %
A	83.0	24	2.4	6.5	31.9
B	83.8	22	3.0	5.7	29.1
C	83.7	26	2.3	5.0	32.2
D	82.0	22	2.8	4.1	32.5
E	82.0	22	2.7	2.2	32.1
F	83.2	24	2.5	1.9	30.2
G	80.0	22	2.4	1.3	32.5

V+E = Vitrinite + Exinite   VM = Volatile Matter

Figure 12 clearly shows the effect of iron sulfide content of the coal on total conversion and liquid product yield during hydrogenation. The conversion increased from about 52 per cent to 70 per cent using the 'hot-rod' reactor with no added catalyst. The yield of toluene soluble product (oil plus asphaltene) increased from about 30 to 44 per cent with total sulfur increase from 1 to 6.5 per cent. Thus it would appear that iron sulfide can act catalytically in the 'dry' hydrogenation reaction as well as in slurried reactions (15).

The iron sulphide in South African coals is a mixture of pyrite and marcasite (18). Although marcasite is known to transform into pyrite at elevated temperatures, separate spiking experiments were performed to see if pyrite or marcasite would show a preferential catalytic effect. The addition of pyrite and marcasite minerals (-200 mesh), to the coal showed equivalent total conversions, and yields of oil and asphaltene.

CONCLUSIONS

For the South African bituminous coals studied here the following conclusions can be made:

  (i)   Conversion yields obtained from coal liquefaction under 'dry' hydrogenation conditions and in the presence of anthracene oil both show good correlations with H/C atomic ratio, the volatile matter yield and the 'reactive' maceral content of the coals.

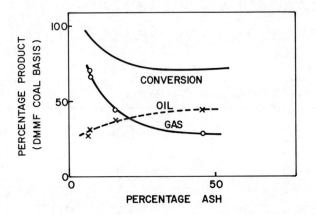

Figure 11.   Product distribution vs. ash content of coal (hot rod mode): catalyst = 1% Sn; P = 25 MPa; T = 500°C.

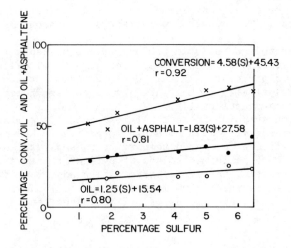

Figure 12.   Effect of sulfur content on liquid yields and overall conversion (hot rod reactor): sand:coal = 2:1; T = 450°C; P = 25 MPa.

(ii)   'Reactive' macerals cannot be defined as the sum of vitrinite + exinite for these South African coals, but substantial portions of the semi-fusinite must be added to obtain total 'reactives'.

(iii)   For 'dry' hydrogenation, good correlations are obtained between toluene soluble yields and the H/C atomic ratio and the volatile matter yield of the coals.

(iv)   The iron sulfide in the coal appears to act beneficially in the 'dry' hydrogenation reaction and enhances the overall liquid yield. No difference was detected in the reactivity of pyrite and marcasite during 'dry' hydrogenation.

LITERATURE CITED

1.   Mackowsky, M-Th. "Coal and Coal Bearing Strata". Eds. Murchison, D.G. and Westroll, T.S., Olivier and Boyd, London, 1968.
2.   Report of the Commission of Inquiry into the Coal Resources of South Africa. Department of Mines, Pretoria, South Africa, 1975.
3.   Hiteshue, R.W., Friedman, S., Madden, R. United States Bureau of Mines, Report of Investigations, 6125, 1962.
4.   Gray, D. Fuel, 1978, 57, 213.
5.   Given, P.H., Cronauer, D.C., Spackman, W., Lovell, H.L., Davis, A., Biswas, B. Fuel, 1975, 54, 40.
6.   Smith, W.H. ISCOR, Personal Communication.
7.   Steffgen, F.W., Schroeder, K.T., Bockrath, B.C. Anal. Chem., 1979, 51, 1164.
8.   Cudmore, J.F. Coal Borehole Evaluation Symposium, Australian Institute of Mining and Metallurgy, 1977, Oct., 146.
9.   Ammosov, I.L., Erexmin, I.V., Sukhenko, S.E., Oshurkova, L.S. Koks i Khimiya, 1957, 12, 9.
10.   Taylor, G.H. Fuel, 1957, 36, 221.
11.   Taylor, G.H., Mackowsky, M-Th., Alpern, B. Fuel, 1967, 46, 431.
12.   Steyn, J.G.D., Smith, W.H. Coal, Gold and Base Minerals (South Africa), 1977, Sept., 107.
13.   Guyot, R.E., Diessel, C.F.K. Australian Coal Industry Research Laboratories, Published Report 79-3, 1978, Dec.
14.   Shibaoka, M., Ueda, S. Fuel, 1978, 57, 667.
15.   Guin, J.A., Tarrer, A.R., Prather, J.W., Johnson, D.R., Lee, J.M. Ind. Eng. Chem. Process Des. Dev., 1978, 17, (2), 118.
16.   Mukherjee, D.K., Chowdhury, P.B. Fuel, 1976, 55, 4.
17.   Gray, D., Barrass, G., Jezko, J., Kershaw, J.R. Fuel, 1980, 59, 146.
18.   Gaigher, J.L. Fuel Research Institute of South Africa, Personal Communication.

RECEIVED March 28, 1980.

The Characteristics of Australian Coals and Their Implications in Coal Liquefaction

R. A. DURIE

R. W. Miller & Co., Pty. Ltd., 213 Miller Street, North Sydney, New South Wales, 2060

In Australia, coal represents, in energy terms, over 97% of the country's non-renewable fossil fuel based energy resources, yet indigenous oil which barely representa 1% of these resources, together with imported oil, supply over 50% of the energy demand with much of this from the transport sector. This situation, catalyzed by the OPEC oil embargo in 1973, has led to strong and sustained interest in the prospects for producing liquid fuels from the abundant coal resources. The reserves of recoverable fossil fuels (1) and the present pattern of energy demand in Australia (2) are shown in more detail in Tables 1 and 2, respectively.

Location, Geology and General Characteristics of Australian Coals

The geographical distribution of Australia's coal resources is shown in Fig. 1. New South Wales and Queensland possess large reserves of black coals in the Sydney and Bowen Basins, respectively, adjacent to the eastern seaboard. Significant deposits of bituminous coals are also known to occur in remote areas in South Australia at Lake Phillipson in the Arckaringa Basin and at currently inaccessible depth (200-300 m) in the Cooper Basin (3,4). [An estimated 3.6×10^6 million tonnes in the latter].

Large reserves of brown coals occur in Victoria with smaller deposits in New South Wales and South Australia.

Whereas the majority of the black coals in the northern hemisphere, including the USA and Europe, were formed during the Carboniferous age, the black coals of Australia are, in the main, Permian. The latter include the coals from the two major basins - the Sydney and the Bowen - and also large deposits in the Galilee Basin (Queensland), at Oaklands (N.S.W.), Lake Phillipson (South Australia) and Collie (West Australia) as well as the deep coal in the Cooper Basin (the Cooper Basin is in the N.E. corner of South Australia extending into the S.W. corner of Queensland (refer Fig. 1)).

0–8412–0587–6/80/47–139–053$05.25/0

Table I   Australia's Fossil Fuel Energy Resources

Resource	Quantity (Demonstrated)	Specific Energy (10^{18} Joules)	(Percentage)
Black Coal*			
In-Situ	48.55×10^9 t	1390	75.5
Recoverable	27.22×10^9 t	780	65.8
Brown Coal			
In-Situ	40.93×10^9 t	400	21.7
Recoverable	39.00×10^9 t	380	32.1
Crude Oil and Condensate			
In-Situ	49.00×10^9 bbl	29.7	1.6
Recoverable	20.70×10^9 bbl	12.4	1.0
Natural Gas + LPG			
In-Situ	545×10^9 m^3	21.0	1.2
Recoverable	327×10^9 m^3	12.6	1.1
Total			
In-Situ		1840.7	100
Recoverable		1185.0	100

*Demonstrated + Inferred in-situ black coal resources are estimated to be 5600×10^{18} J with 55% recoverable - inferred resources of crude oil and natural gas are relatively minor representing only 1% and 8%, respectively, of the demonstrated resources.

Table II   Pattern of Australian Use of Fossil Fuels 1974-75

Total primary energy demand 2512×10^{15} J consisting of:
coal 1035×10^{15} J; oil 1318×10^{15} J; natural gas 159×10^{15} J

	% of Fuel Type	% Total Primary Energy
Coal		
Electricity generation	61	26)
Iron and steel	25	10) 42
Other	14	6)
Oil		
Transport*	61	32)
Fuel oil	15	8) 52
Other	24	12)
Natural Gas		
Electricity generation	20	1
Other	80	5) 6

*Includes fuel oil for bunkering

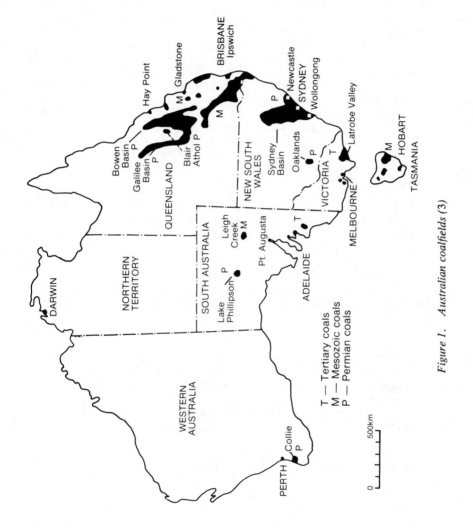

Figure 1. Australian coalfields (3)

These Permian coals, together with counterparts in India, South
Africa, Antarctica and South America, are referred to as Gondwana
coals after the hypothetical super-continent which subsequently
broke up into the continents and sub-continents mentioned above (5).

The climatic conditions prevailing in the Permian during the
formation of these Gondwana coals were different from those for
the Carboniferous coals of North America and Europe. As a result
of a cooler climate with alternating dry and wet periods, and of
the consequent difference in the original plant materials, the
conditions of accumulation, the slower rate of accumulation, and
prolonged duration of sinking, the Australian (and other Gondwana)
Permian coals differ in many respects from the Carboniferous coals
of the northern hemisphere. Thus for the former coals, seam thick-
ness tends to be greater, vitrinite content lower, semi-fusinite
content higher, mineral matter content high and sulphur content
generally low; the ash derived from the mineral matter is usually
refractory with high fusion temperatures. These coals occur in
seams near the surface, and at depth.

The Australian Permian coals vary widely in rank (maturity)
and type (vitrinite content) from the Oaklands (N.S.W.) coal at
72% (dry ash-free basis) carbon, a hard brown coal (6), containing
17% vitrinite, at one extreme - through high volatile bituminous
coals such as Galilee (Queensland) coal at 77% carbon, 16% vitrinite; Blair Athol (Queensland) coal at 82% carbon, 28% vitrinite,
Liddell (N.S.W.) coal at 82% carbon, and >70% vitrinite - to low
volatile bituminous such as Peak Downs (Queensland) at 89% carbon,
71% vitrinite, and Bulli seam (N.S.W.) 89% carbon, 45% vitrinite.

In addition to the Permian coals there are occurrences of
Mesozoic and Tertiary coals in Australia. Mesozoic coals occur in
small basins in South Australia, Tasmania, New South Wales and
Queensland and vary in rank from brown to bituminous. Perhaps the
most notable occurrences in the present context are the Walloon
coals in the Clarence-Morton basin in Queensland, e.g. Millmerran
bituminous coal (78% carbon, vitrinite plus exinite ~90%).

The most significant Tertiary coals are represented by the
vast brown coal deposits in Victoria, particularly in the Latrobe
Valley. These brown coals with 68-70% carbon, occur in very thick
seams (up to 200 meters) under shallow cover (<30 meters). These
coals differ from the Tertiary brown coals of North America in
that they have a much lower ash yield and significant amounts of
the ash-forming inorganic constituents are present as cations on
the carboxylic acid groups which are a characteristic of low rank
coals.

Coal Characteristics and Their Effects in Liquefaction Process
The wide variation in Australian coals in rank, type and inorganic impurities and the significant differences between these

coals and those from the USA and elsewhere, emphasize the need for
detailed understanding of how specific coal characteristics influence liquefaction reactions and the properties of the liquid
product. The heterogeneity and variability of coals make them a
complex feedstock and presents major challenges to efforts to
identify and quantify those parameters of most significance. However, until this is achieved the application of a process developed
and optimized on a coal, or similar coals, from one region to
coals in another region is fraught with danger. In recognition of
this, research is in progress in a number of laboratories in
Australia to elucidate the chemistry of Australian coals in
relation to their liquefaction. This encompasses both black and
brown coals and liquefaction via pyrolysis, non-catalytic hydrogenation (solvent refining) and catalytic hydrogenation. The
results obtained in these studies are informative but some give
rise to more questions than answers. In the remainder of this
paper selected highlights from these Australian studies will be
presented and discussed.

Effects of Petrographic Composition and Rank

It is possible to produce some liquid hydrocarbons from most
coals during conversion (pyrolysis and hydrogenation, catalytic
and via solvent refining), but the yield and hydrogen consumption
required to achieve this yield can vary widely from coal to coal.
The weight of data in the literature indicate that the liquid
hydrocarbons are derived from the so-called 'reactive' macerals,
i.e. the vitrinites and exinites present (7,8,19). Thus, for
coals of the same rank the yield of liquids during conversion
would be expected to vary with the vitrinite plus exinite contents.
This leads to the general question of effect of rank on the
response of a vitrinite and on the yield of liquid products; and,
in the context of Australian bituminous coals, where semi-fusinite
is usually abundant, of the role of this maceral in conversion.

A number of research projects in Australia are being addressed to these questions. The Australian Coal Industry Research
Laboratories (ACIRL) have been approaching the question through a
study of the conversion behaviour of a selected range of Australian bituminous coals under non-catalytic solvent refining
conditions (9,10). The Commonwealth Scientific and Industrial
Research Organization (CSIRO) is considering the question with
regard to pyrolysis (11) and catalytic hydrogenation (13) of
bituminous and brown coals, with support from studies of the
behaviour of individual maceral types during conversion with the
aid of petrographic techniques (12).

Experimental data published recently by Cudmore (10) for
eight Australian bituminous coals, reproduced in Fig. 2, show a
direct linear correlation between conversion (to gas + liquids),
under non-catalytic hydrogenation conditions using Tetralin as

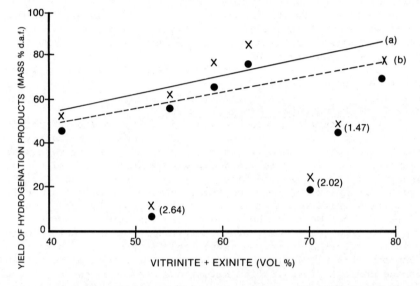

Figure 2. Noncatalytic hydrogenation—product yields vs. vitrinite + exinite content. Curve a, total conversion (×); Curve b, extract (●) (10). (Note: Lines a and b relate to coals where the mean maximum reflectance (Rₒ max) of the vitrinite fall in the range 0.43–0.68%. Values in parenthesis refer to Rₒ max for higher-rank coals.)

vehicle, and the vitrinite plus exinite contents over the range 40
to 80%, for coals in the rank range where the mean maximum reflectance ($\bar{R}_O$ max.) of the vitrinite varies from 0.43 to 0.68%, i.e.
for carbon content over the range of about 75% (dry ash-free) to
about 82%. This encompasses the sub-bituminous coals and high
volatile bituminous coals. However, for coals where $\bar{R}_O$ max. is
greater than 1.47% the yield was markedly lower than might otherwise have been expected from the vitrinite plus exinite contents
(refer Fig. 2). The information in Fig. 2 would further suggest
that the rank effect in decreasing conversion yield increases
rapidly with increase in rank from $\bar{R}_O$ max. 1.47% to 2.64%, i.e.
carbon (dry ash-free basis) 88% to >90%. This, of course, leaves
open the question of where the decrease in the conversion of the
vitrinite (+ exinite) starts in the rank range 83 to 88% carbon.

An implication of Cudmore's data ([10]) for the sub-bituminous
and high volatile bituminous coals is that the semi-fusinite as
such appears to contribute little to the conversion products,
otherwise the apparent dependence of yield on the vitrinite
(+ exinite) content would not be so linear.

The whole question relating to the possible role of semi-
fusinite is receiving the attention of Shibaoka and his associates
in CSIRO ([12]). Although the project is still at an early stage,
direct observations on the changes occurring in semifusinite-rich
coal grains during conversion under a wide variety of conditions
suggest that the possible contributions of this maceral in conversion cannot be ignored although further work is required to
define the nature and magnitude of such contributions.

Studies initiated by the author in CSIRO ([13]) seek to throw
light on the role of the various macerals by studying the conversion, under catalytic hydrogenation conditions, in Tetralin as
vehicle, of maceral concentrates from a high volatile bituminous
coal. Some preliminary results, given in Fig. 3, show conversions
as almost complete for the hand picked vitrain (>90% vitrinite)
from a high volatile bituminous coal (Liddell seam N.S.W., 83.6%
carbon and 43% volatile matter both expressed on a dry ash-free
basis). However, it is evident that the conversion of the 'whole'
coal increases rapidly with increase in hydrogen pressure (under
otherwise similar conditions - batch autoclave, 4h. @ 400°C).
This could suggest either that conversion of the vitrinite is suppressed by other components in the coal, particularly at the lower
pressures, or more likely, that other macerals are participating
to an increasing extent as the hydrogen pressure increases.

Consideration of the latter results in relation to those of
Cudmore ([10]), discussed above, emphasize the need for caution when
generalising on the influence of coal characteristics on conversion.
Indeed, it would appear that the absolute and relative

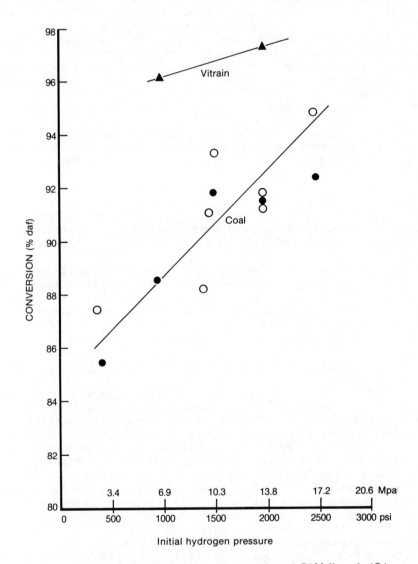

Figure 3.   Effect of hydrogen pressure on conversion of Liddell coal: (○), un-treated coal; (●) demineralized coal; (▲), hand-picked vitrain. Reaction tempera-ture = 400°C; reaction time = 4 hr.

contribution of the various petrographic components is dependent on the process conditions which include, inter alia, the hydrogen potential.

The petrography of brown coals differs from that of black coals and is less well developed. However, evidence is mounting that brown coals can vary significantly, even within the same seam, and that these variations may effect their conversion behaviour. The Victorian Brown Coal Council has initiated studies in this area (with advice from the German Democratic Republic).

The Effect of Elemental Composition
   It is well established that for any coal the so-called re-active macerals, vitrinite and exinite, are richer in hydrogen than the inert macerals. Therefore, since the conversion of coals to liquid fuels involves the production of lower molecular weight products having atomic hydrogen to carbon ratios in the range 1.7 to 2 compared with <1 for most coals, it is of interest to consider the effect of the hydrogen content, or alternatively the hydrogen/carbon ratio on the conversion of coals to liquid and gaseous fuels under a wide range of conditions.

   Pyrolysis. In this context it is relevant to consider initially the effect of hydrogen contents on tar yields during pyrolysis (carbonization). This is particularly so, since, in all coal conversion processes little happens until the coal is at a temperature above that where active thermal decomposition normally sets in. In other words, all coal conversion processes may be regarded as pyrolysis under a variety of conditions which determine the nature of the primary decomposition and the reactions which follow.

   Fig. 4 represents a plot of the atomic H/C ratio versus tar yields obtained by the former CSIRO Division of Coal Research for a wide variety of Australian coals during low temperature (600°C) Gray-King carbonization assays (14) over several years. This figure shows that, despite a wide variation in rank and inorganic impurities, there is a significant linear correlation between the tar yield and the atomic H/C ratio. A variety of factors may account for the scatter - the empirical nature of the assay, wide variations in the ash yield and nature of the ash (see below), weathering of the coal, the multitude of analyses involved and the long time span over which the results were accumulated.

   The steep dependence on hydrogen content of the tar yields obtained during the low temperature (500°C) fluidized bed carbon-ization of 14 Australian coals, ranging in rank from 72% to ~89% (dry ash-free basis) carbon content, is clearly demonstrated in Fig. 5 (15,16).

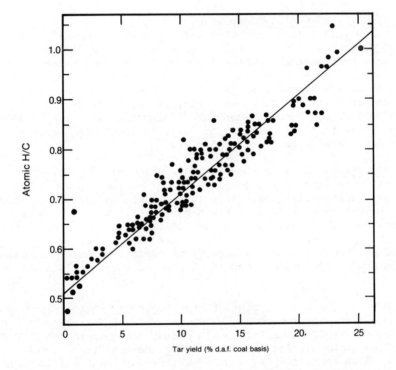

Figure 4.   Dependence of tar yield, determined by low-temperature Gray–King carbonization assay, n atomic hydrogen-to-carbon ratio for a wide range of Australian coals.   Tar yield = 50.4 × H/C − 25.9; correlation coefficient, 0.95.

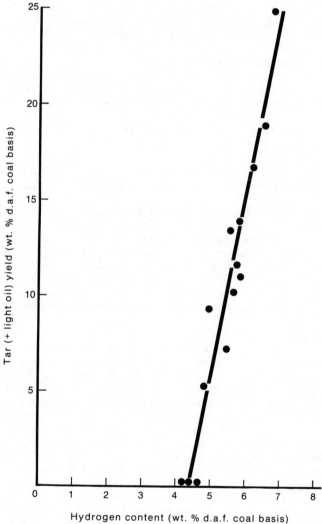

Figure 5.   Dependence of tar yields from low-temperature (500°C), fluidized-bed carbonization of hydrogen content for some Australian coals (15, 16)

In current CSIRO investigations into the production of liquid fuels via the flash pyrolysis of selected Australian coals (11) the importance of the hydrogen content, or more precisely the atomic H/C ratio of the coal with regard to the total yield of volatile matter and tar, has been demonstrated also. This is shown in Fig. 6 (20) together with the reproduction of the correlation line for the low temperature Gray-King Carbonization assay transferred from Fig. 4. Also included are data obtained for one USA bituminous coal (Pittsburgh No. 8) and one lignite (Montana). The former coal plots consistently with the Australian bituminous coals for both the volatile matter and tar yield; but whereas the raw Montana lignite, together with the raw Australian brown coal, are consistent with the bituminous coals for total volatile matter yield, the tar yields from the lignite and brown coal fall significantly below those to be expected from Fig. 4 for bituminous coal with similar atomic H/C ratios with one exception – a low ash sample of Loy Yang brown coal. The reason for the 'deviation' is considered in the next section of the paper.

Hydrogenation. Cudmore (10) in his studies of the non-catalytic hydrogenation (solvent refining) of six Australian coals has indicated that the conversion systematically increases as the atomic H/C ratio of the coal increases over the range 0.60 to 0.85. This is shown in Fig. 7 (10) which also includes data for the catalytic hydrogenation of six Canadian coals (17). These results, together, indicate the importance of the hydrogen contents of coal in general for both non-catalytic and catalytic hydrogenation.

With regard to the implications of the elemental composition (ultimate analysis) of Australian coals, brown coals (lignites) call for special attention by virtue of their high oxygen contents (as high as 30%). During hydrogenation of brown coals it is usually considered that significant amounts of hydrogen are consumed in the elimination of oxygen as water and that this places these coals at a disadvantage because the cost of hydrogen is a significant factor in the economics of conversion. White has recently considered oxygen balances in the catalytic hydrogenation of some Australian brown coals (18). This study indicates that, whereas the overall conversion, under comparable conditions, is higher for brown coals than for bituminous coals studied, the yield of hydrocarbon liquids is higher for the latter; but, surprisingly, the hydrogen consumption in the primary conversion is actually lower for the brown coals than for any of the bituminous coals studied (H/C in range 0.57 to 0.72). Further, the results show that the percentage hydrocarbon liquid yield (dry ash-free basis) per percent of hydrogen consumed is actually as high, or higher, for a low ash yield (0.5%) brown coal (H/C = 0.81) by comparison to the bituminous coals studied. Indeed, the evidence suggests that much of the oxygen (~30%) is eliminated as carbon monoxide and carbon dioxide).

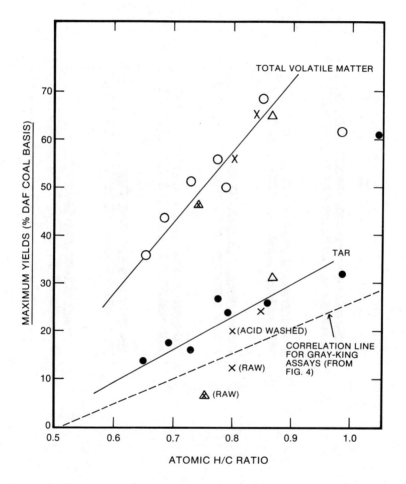

Figure 6.   Dependence of maximum tar yields and corresponding total volatile matter yields during flash pyrolysis on atomic hydrogen-to-carbon ratio for some Australian and U.S.A. coals: (○, ●), black coals; (×), brown coals; (▲), Pittsburgh No. 8 (U.S.A.); (⬭), Montana lignite (U.S.A.).

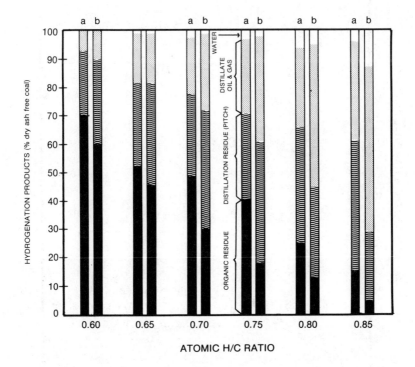

Fuel Processing Technology

Figure 7. Dependence of yields of hydrogenation products on the atomic hydro-gen-to-carbon ratio: (a) Australian coals—noncatalytic conditions (10); (b) Cana-dian coals—catalytic conditions.

Effect of Inorganic Constituents

Despite much speculation on the possible effects of the inorganic ash-forming constituents in a coal on its behaviour during
conversion, there is still no clear understanding on the subject.
It is generally suspected that where pyrite is present in a coal
this is converted to pyrrhotite under the conditions of coal
hydrogenation and can act as a catalyst (19). The effectiveness
will, of course, be dependent on how the pyrite is disseminated
through the coal including the maceral association; this may be
the cause where no significant effect has been noted (8). In the
majority of Australian coals the sulphur, and hence pyrite, content
is very low and hence the possibility of a catalytic effect from
pyrite is negligible. As mentioned earlier, Australian bituminous
coals tend to be high in mineral matter. This consists primarily
of alumino-silicate minerals (20). To prepare most coals for use
as a feedstock in conversion these will need to be processed in a
coal preparation plant to reduce the ash yield. Otherwise reactor
throughput in terms of effective coal feed rates, are adversely
affected and excessive ash can 'blind' added catalysts and cause
other operational problems. Since alumino-silicates are the basis
of cracking catalysts, the mineral matter in the coal might well
act in this way and be either to the advantage or disadvantage of
the conversion process.

A project initiated by the author when with CSIRO has, as one
of the objectives, the study of effect of the mineral matter in
selected Australian coals during catalytic hydrogenation (13).
The initial approach has been to compare, under otherwise identical
conditions, the conversion behaviour of a coal sample before and
after demineralization. Some very preliminary results are shown
in Fig. 8 for a sample of Liddell seam coal (ash yield 7.35% air
dried basis; volatile matter 43.2% and total sulphur 0.48% dry
ash-free basis) before and after demineralization to reduce the
ash yield to 0.5%. Fig. 8 shows the effect of temperature on the
total conversion and yield of bitumen, (i.e. residue from
atmospheric and vacumn distillation to 210°C of the hydrogenation product) during batch catalytic (a commercial Co-Mo on
alumina catalyst) hydrogenation using Tetralin as solvent. The
main effect of the mineral matter appears to be to give an
increased scatter in the experimental data with regard to total
conversion. This is also evident, but to a lesser degree, in
Fig. 3 where the effect of initial hydrogen pressure on total conversion for the same coal is indicated.

Since the scatter of experimental points for total conversion
is both above and below the curve for the demineralized sample, it
is not possible to assign the behaviour of the untreated coal to
either catalyst blinding or enhanced catalytic effects. With
regard to the yield of 'bitumen' (Fig. 8), the bias on the high
side in yield could be interpreted to suggest that some catalytic

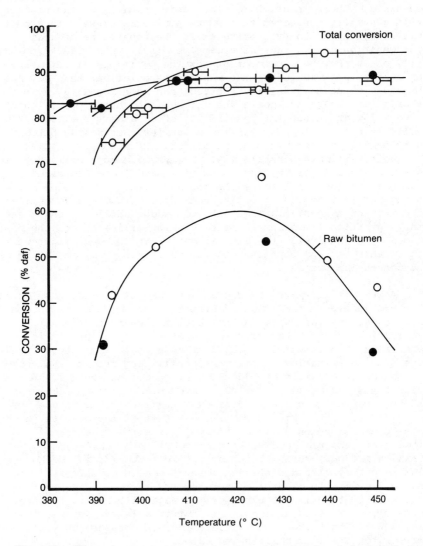

Figure 8. Effect of temperature on conversion of untreated and demineralized Liddell coal (300-mL autoclave, 6.9 MPa, 4 hr): (○), untreated; (●), demineralized.

effect was exhibited by the mineral matter. Obviously further studies of this type are required to determine the nature of the effects, if any, of the alumino-silicates in Australian bituminous coals on the response of these coals during conversion.

Australian brown coals are of special interest with regard to the possible influence of the inorganic constituents during pyrolysis and hydrogenation. In the low ash yield Australian brown coals, a considerable proportion of the inorganic ash-forming constituents are present as cations associated with the carboxylic acid groups in the coal (21,22,23). Studies in CSIRO have shown that the nature and amount of these cations can exert marked effects on the behaviour of the coal during thermal decomposition (pyrolysis). In particular, Schafer (24) has shown that the presence of cations facilitate the elimination of the oxygen during pyrolysis in a manner that is still not understood. This could have interesting and practical opportunities for upgrading brown coal as a feedstock for hydrogenation.

In the USA, observations with North Dakota lignites have suggested that sodium associated with the carboxyl groups have a beneficial catalytic effect with regard to the quality of the liquid product (8). Further, the superiority of CO-steam over hydrogen in the 'non-catalytic' hydrogenation of lignite has been attributed again to the catalytic effects of alkali and alkaline earth metals present on the coal (25) which are known to be effective catalysts in the carbon-steam and carbon monoxide-steam reactions. It has been suggested that the hydrogen generated accordingly in-situ is more effective since it probably passes transiently through the reactive 'nascent' hydrogen-form and avoids the need to dissociate the strong bond in molecular hydrogen.

The ability to exchange cations on the carboxylic acid groups in brown coal (26) has led to interest into the effectiveness of transition metals exchanged onto the carboxyl groups as catalysts. This aspect was first looked at by Severson and his colleagues in North Dakota with negative results (27). However, the matter is now being re-examined in Australia in the context of Victorian brown coals. Careful studies in this area could well help contribute to the better understanding of the role of the catalyst in coal hydrogenation, e.g. does it facilitate the direct transfer of hydrogen from molecular hydrogen in the gas phase, or in solution, to the fragments derived from the thermally decomposing coal? or does it simply facilitate in the regeneration of the hydrogen donor capacity of the 'solvent'?

It is appropriate to conclude this section by reference to one aspect of the CSIRO flash pyrolysis project involving, again, brown coals. Here, it has been shown (28) that the presence of

cations on the carboxyl groups strongly supresses the tar yield
obtained during rapid pyrolysis. For example, a sample of raw
Gelliondale (Victoria) brown coal having a 7.2% (dry basis) ash
yield, yielded 12% (dry ash-free basis) of tar during flash
pyrolysis but, when this coal was acid washed 0.7% (dry basis) ash
yield, the tar yield increased to 20% (dry ash-free basis).
Further reference to Fig. 6 shows that the latter tar yield now
plots with the bituminous coals with reference to the effect of
the atomic H/C ratio. Similarly a second brown coal sample (Loy
Yang) which, as recovered from the seam, has a very low ash yield
(0.4% dry ash-free basis), and most of the carboxyl groups in the
acid form, plots with the bituminous coals in Fig. 6; however,
when the sodium-salt is produced from this coal before flash
pyrolysis the tar yield is almost complete supressed.

It is interesting to speculate on the significance of these
observed effects of the presence of cations on the carboxylic
acid groups in brown coals. It would appear that the cations
either inhibit the tar forming reactions in some way or else cause
the tars, once formed, to polymerize to a solid residue. The
former possibility could imply that the tars are formed from lower
molecular weight precursors by reactions which are blocked by the
presence of a cation, or cations, on the carboxyl groups and the
latter that these cations inhibit the escape of the tars. The
clue to the detailed explanation perhaps resides with the observations, already mentioned, of Schafer (24) on the effects of
cations associated with carboxyl groups on the oxygen elimination
reactions during the thermal decomposition of brown coals.

Further detailed studies in this area are obviously needed
to resolve the chemistry involved. Such pyrolysis studies sup-
plemented by hydrogenation experiments with acid-form and salt-
form brown coals offer promise of resolving the precise role of
pyrolysis in the hydrogenation of these coals and of how the
ash-forming cations participate in the hydrogenation reactions.
For example, how does the presence of the cations effect the
hydrogen consumption? A question that needs also to be considered
in the context of the observations of White (18).

CONCLUDING REMARKS

The first part of this paper has shown that Australian black
and brown coals differ significantly in a number of respects from
coals of similar ranks from North America and elsewhere in the
northern hemisphere. The rest of the paper than proceeded to
indicate the progress being made to determine how the characteristics of Australian coals influence their conversion to volatile
and liquid products during pyrolysis and hydrogenation.

The results presented and discussed here for current investigations on Australian black coals indicate strongly that,

over a rank range up to about 83% (dry ash-free) carbon, the vitrinite and exinite contents and overall, the atomic hydrogen-to-carbon ratio are the important parameters with regard to total volatile and liquid yields during pyrolysis and hydrogenation of such coals. In these respects there appears to be no major differences relative to northern hemisphere coals. The strong dependence of conversion on atomic H/C ratio suggest that the subleties of variation in chemical composition or structure with change in rank are of secondary importance. Also the near linear dependence of conversion yields on the atomic H/C ratio further suggest that the effects of the mineral matter in the Australian black coals may be secondary.

The results mentioned for Australian brown coals raise many interesting questions concerning the effect of coal characteristics on conversion during pyrolysis and hydrogenation. These relate to the similarity of the behaviour of the acid-form brown coals with the black coals in terms of the effect of the atomic H/C ratio on conversion during pyrolysis; the suppression of the tar yield when the carboxyl groups are in the salt-form; and the elimination of oxygen during the primary hydrogenation without the involvement of hydrogen. Again, within the limitations of the investigations mentioned, there is no reason to believe that the effects observed should be unique to Australian brown coals.

It is emphasised that many of the results discussed relate to on-going investigations and need confirmation on other coals. Also, many of the effects mentioned relate to the overall conversion. In coming to grips with the effects of coal characteristics, attention must be given to the quality as well as the quantity of liquid products obtained during conversion; as well as to the rate at which the conversion occurs under various conditions. These aspects, which have not been considered in this presentation, call for careful experimentation where the emphasis is not on maximising conversion but on careful control of experimental conditions with termination of experiments at only partial conversion.

ACKNOWLEDGEMENTS

The author gratefully acknowledges the co-operation and help he has received from the Australian Coal Industry Laboratories (ACIRL), the Melbourne Research Laboratories of the Broken Hill Proprietary Co. Ltd. (MRL/BHP), the Commonwealth Scientific and Industrial Research Organization (CSIRO), in providing information and data, often unpublished, to assist in the preparation of this paper. In particular, he wishes to thank Dr. N. White (MRL/BHP), Mr. J. Cudmore (ACIRL), and the following former colleagues in CSIRO - Prof. A.V. Bradshaw, Dr. D. Jones, Mr. H. Rottendorf, Mr. H.N.S. Schafer, Dr. M. Shibaoka, Mr. I.W. Smith, Dr. G.H. Taylor, and Mr. R.J. Tyler.

REFERENCES

1. National Energy Advisory Committee (Commonwealth of Australia),
    "Australia's Energy Resources : An Assessment", Report No.2,
    (published by Dept. of National Development), December 1977.

2. Durie, R.A., "Coal Conversion Research in Australia", Paper B1,
    Third International Conf. on Coal Research, Sydney, Aust.,
    October 1976.

3. Taylor, G.H. and Shibaoka, M., "The Rational Use of Australia's
    Coal Resources", Paper 8, Institute of Fuel (Australian Mem-
    bership) Conference on "Energy Management", Sydney, Aust.,
    November 1976.

4. Australian Petroleum Institute, Petroleum Gazette, March 1978,
    p.12.

5. Stach, E.; Mackowsky, M,Th.; Teichmüller, M.; Taylor, G.H.;
    Chandra, D.; Teichmüller, R., "Coal Petrology", Borntraeger,
    Berlin 1975.

6. International Committee for Coal Petrology, 2nd Edition, Paris,
    Centre National de la Recherche Scientifique, 1963.

7. Fisher, C.H.; Sprunk, G.C.; Eisner, A.; O'Donnell, H.J.;
    Clarke, L.; Storch, H.H., "Hydrogenation and Liquefaction of
    Coal, Part 2 - Effect of Petrographic Composition and Rank
    of Coal", Technical Paper 642, US Bureau of Mines, 1942.

8. Given, P.H.; Cronauer, D.C.; Spackmen, W.; Lovell, H.L.;
    Davis, A.; Biswas, B., Fuel, 54, 34, 40, 1975.

9. Guyot, R.E., "Influence of Coal Characteristics on the Yields
    and Properties of Hydrogenation Products", Aust. Coal
    Industries Research Laboratories, Report PR 78-8, June 1978.

10. Cudmore, J.F., Fuel Processing Technology, 1, 227, 1978.

11. Tyler, R.J., Fuel (Ldn.), in press, 1980.

12. Taylor, G.H.; Shibaoka, M., and Ueda, S., CSIRO Fuel Geo-
    sciences Unit, private communication.

13. Jones, D.; Rottendorf, H.; Wilson, M., CSIRO Division of
    Process Technology, unpublished results.

14. Durie, R.A. and Shibaoka, M., unpublished results.

15. Bowling, K.McG. and Waters, P.L., CSIRO Division of Coal Research, Investigation Report 74, September 1968.

16. Kirov, N.Y. and Maher, T.P., Paper 11, Institute of Fuel (Australian Membership) Conference on "The Changing Technology of Fuel", Adelaide, November 1974.

17. Boomer, E.H.; Saddington, A.W.; Edwards, J., Canadian J. Res. 13B, 11, 1935.

18. White, N., The Broken Hill Proprietary Co. Ltd., Melbourne Research Laboratories, private communication.

19. Mukhergee, D.K.; Sama, J.K.; Choudhury, P.B.; Lahiri, A., Proc. Symposium Chemicals and Oil from Coal, CFRI India, 1972.

20. Brown, H.R. and Swaine, D.J., J. Inst. Fuel, 37, 422, 1964.

21. Durie, R.A., Fuel (Ldn.), 40, 407, 1961.

22. Baragwanath, G.F., Proc. Aust. I.M.M. (202), 131, 1962.

23. Burns, M.S.; Durie, R.A.; Swaine, D.J., Fuel (Ldn.), 41, 373, 1962.

24. Schafer, H.N.S., Fuel (Ldn.), 58, 667, 1979.

25. Appell, H.R.; Wender, I.; Miller, R.D., U.S. Bureau of Mines, I.E. 8543, 1972.

26. Schafer, H.N.S., Fuel (Ldn.), 49, 271, 1970.

27. Severson, D.E.; Souby, A.M.; Kuban, W.R., Proc. 1973 North Dakota Symposium, U.S. Bureau of Mines, I.C. 8650, 1974.

28. Schafer, H.N.S. and Tyler, R.S., Fuel (Ldn.), in press.

RECEIVED March 28, 1980.

Relationship Between Coal Characteristics and Its Reactivity on Hydroliquefaction

K. MORI, M. TANIUCHI, A. KAWASHIMA, O. OKUMA, and T. TAKAHASHI

Mechanical Engineering Laboratory, Kobe Steel, Ltd., Iwaya, Naka-ku, Kobe 657, Japan

It has recently been acknowledged that in future coal will play a more important role as an energy source for petroleum. Especially in Japan, whose energy sources depend largely on imported petroleum, the development of coal technology must be accelerated to prepare against a future energy crisis. Coal liquefaction, one of the processes that promises to solve this crisis, is now in the development stage. As with petroleum, Japan depends on imported foreign coal, because of its own peculiar coal mining conditions. But in Japan, a wide variety of coal species will be used for liquefaction. Therefore, the effect of characteristics of coal on reactivity during liquefaction is an important research subject for selecting the coal species.

Location, Geology and General Characteristics of Japanese Coals

The geographical distribution of Japan's main coal fields and coal mines is shown in Fig.1. Though Japan is composed of four main islands, i.e., Hokkaido, Honshu, Shikoku and Kyushu, from the north to the south, the coal resources are mainly limited to Hokkaido and Kyushu as shown in Table 1. Although the majority of the Japanese coals were formed during the Cenozoic era in the Tertiary period, their coalifications are extraordinarily advanced owing to the crustal movements and volcanic activities they have experienced; therefore Japan produces a wide range of coals varying from brown coal to anthracite.

The properties of Japanese coal and the fields can be characterized, in comparison with those of the continental type, as follows:

(1) Coal fields are small in scale and defective in continuity.

(2) Geological structure is complicated due to numerous faults and foldings.

0–8412–0587–6/80/47–139–075$05.50/0

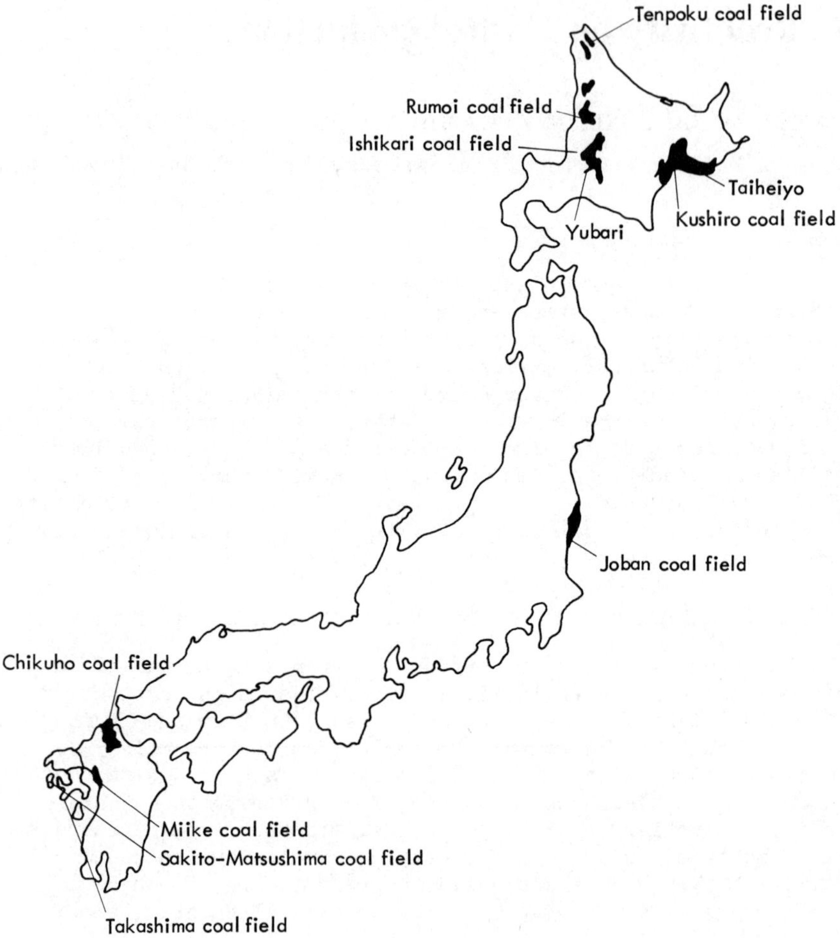

Figure 1. Main coalfields in Japan

Table 1. Japanese coal reserves

(10⁶ tons)

Region	Theoretical Minable Coal Reserves*			Total
	Proven	Indicated	Inferred	
Hokkaido	2,298	1,105	2,950	6,353
Honshu	798	308	660	1,766
Kyushu	2,406	912	2,951	6,269
Japan's Total	5,502	2,325	6,561	14,388

* Coal reserves up to the present depth levels of mining technology.

This data is quoted from the special report, 1973 of Research Coordination Bureau, Sience and Technology Agency.

(3) Coal is rich in hydrogen or volatile matter and higher in heating value.

(4) Caking property is not strong but some are of extremely high fluidity.

Liquefaction Behaviour of Coals

It is well known that the characteristics of coal differ widely according to the age of the coal formation as well as to the location of coal, etc. And the reactivity during hydro-liquefaction depends on the characteristics of coals. This relationship will be a guidance to select and develop coal mines. Many parameters to indicate the reactivity of coal have been proposed (1, 2, 3). Among these parameters, carbon content, volatile matter content, value of H/C atomic ratio, reactive macerals' content, etc. are reported to be relatively closely related parameters to coal reactivity. However, these relations are usually found only in limited reaction conditions. Therefore, attempts to find better parameters still continue.

In this study, we have tried to find a more comprehensive parameter related to coal reactivity, as represented by conversion, by liquefying several ranks of coals. These cover a wide range from lignite to bituminous coal. Also we have studied the difference of coal reactivity caused by the mining sites in Australian brown coal mines. Selected coals from a wide range of rank are located in the coal band shown in Fig.2. The resulting parameters are compared with other parameters reported by other researchers (2, 3).

Experiments and Results

Analytical data on coals used in this study are presented in Tables 2 and 3. Hydroliquefaction data on coals used in this study are summarized in Tables 4 and 5.

The liquefaction of coals was studied in a 500 ml magnetically-stirred stainless steel antoclave. Two different reaction conditions were used in this study, but the experimental procedures were almost the same in both conditions.

Coal, solvent and catalyst were charged to the autoclave. After the autoclave had been flushed and pressurized with hydrogen to the desired initial pressure, the autoclave was heated with constant electric power and with constant stirring up to the reaction temperatures. Then, the autoclave was held at these temperatures for periods of the desired length. At the conclusion of the reaction, the autoclave was quenched by dropping the heating jecket and cooled by standing in air until it reached room temperatures. After cooling, the reaction gases were vented,

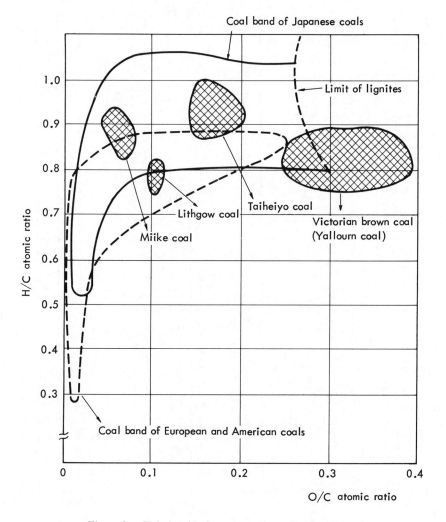

Figure 2. Relationship between coals used and coal bands

Table 2. Analytical data on coals used in the study of the wide range

Coal quality / Coal type	Ultimate analysis (d.a.f.)							Proximate analysis *				VC [5]	Inerts
	C	H	O	N	S	H/C	O/C	M [1]	A [2]	VM [3]	FC [4]		
Yallourn brown coal	62.4	4.2	32.6	0.6	0.2	0.800	0.392	11.4	0.8	45.5	42.3	14.2	
Taiheiyo coal	73.0	5.6	19.4	1.6	0.4	0.912	0.199	6.3	13.6	40.2	39.9	23.2	8.9
Lithgow coal	81.5	5.1	11.0	1.6	0.8	0.744	0.101	1.7	12.1	32.6	53.6	19.3	36.2
Miike coal	81.5	5.7	8.3	1.3	3.2	0.832	0.076	0.7	13.4	39.8	46.1	27.8	3.7

* Equilibrium moisture content,   1) Moisture,   2) Ash,   3) Volatile Matter,
4) Fixed carbon,   5) Volatile carbon.

Table 3. Analytical data on Morwell brown coals used in the study of the narrow range

Coal quality / Coal type	Ultimate analysis (d.a.f.)							Proximate analysis (dry)				V C	Litho-type
	C	H	O	N	S	H/C	O/C	M *	A	VM	FC		
A	67.05	3.70	28.36	0.58	0.31	0.656	0.317	23.3	2.4	49.4	48.2	17.66	Dark
B	69.36	4.12	25.45	0.82	0.26	0.713	0.275	15.2	2.3	45.8	51.9	16.24	Medium Dark
C	69.82	4.45	24.49	0.92	0.32	0.765	0.263	15.7	2.7	47.3	50.0	18.43	Medium Light
D	69.49	4.72	24.79	0.76	0.27	0.815	0.267	14.5	2.9	53.7	43.4	24.79	Light
E	71.85	4.90	21.95	0.87	0.43	0.811	0.229	21.2	2.7	51.2	46.0	24.52	Light

* Equilibrium moisture content

Table 4.   Reaction conditions and results of hydroliquefaction on coals used in the study of the wide range

Feed Coal	Yallourn coal		Lithgow coal		Taiheiyo coal		Miike coal	
Feed Coal (g. as d.a.f.)	43.9		43.1		40.1		43.0	
Solvent*	150		150		150		150	
Catalyst (g) Fe$_2$O$_3$	0	0.75	0	0.75	0	0.75	0	0.75
S	0	0.15	0	0.15	0	0.15	0	0.15
Hydrogen initial pressure (kg/cm^2)	60		60		60		60	
Reaction Temperature (°C)	450		450		450		450	
Holding Time at Reaction Temp. (hr.)								
Conversion** (%)	28.5	55.0	41.8	58.9	60.3	65.2	91.1	97.4

* Solvent consists of creosote oil and recovered solvent.

** Conversion was calculated by benzene insoluble residue.

Table 5.   Reaction conditions and results of hydroliquefaction on Morwell brown coal used in the study of the narrow range

Feed coal sample	A	B	C	D	E
Feed coal (g. as d.a.f.)	37.5	37.5	37.5	37.5	37.5
Solvent*	112.5	112.5	112.5	112.5	112.5
Catalyst (g) Fe$_2$O$_3$	0.54	0.54	0.54	0.54	0.54
S	0.22	0.22	0.22	0.22	0.22
Hydrogen initial pressure (kg/cm^2)	80	80	80	80	80
Reaction Temperature (°C)	430	430	430	430	430
Holding Time at Reaction Temp. (hr.)	1.0	1.0	1.0	1.0	1.0
Conversion**	59.0	75.8	81.6	87.7	80.8

* Solvent means creosote oil.

** Conversion was calculated by pyridine insoluble residue.

and collected in a gas sampling flask. The final products left
in the autoclave were filtered by suction. The residue left on
the filter was transferred to a Soxhlet extractor and extracted
with benzene or pyridine until the washing solvent was a light
yellow color. After extracting, the weight of the insoluble
residue was determined after being dried at 120°C, under 5 mmHg,
and over 2 hrs, using a vacuum-drier. The filtrate from the
reaction mixture and the concentrated solution from the washing
solvent were combined and then vacuum distilled up to 310°C at
90 mmHg. The fractions with boiling point : 120 - 310°C and
310°C above (the vacuum bottom) were recovered as solvent and
SRC, respectively.
Conversion was calculated as follows :

Conversion %

$$= \frac{\text{Coal charged (d.a.f.)} - \text{Insoluble residue (d.a.f.)}}{\text{Coal charged (d.a.f.)}} \times 100$$

     The analytical data for coal samples used by other resear-
chers and their experimental results are shown in Tables 6 and
7. A rough comparison of the liquefaction conditions used in
this study to explore the parameter representing coal characteristics is shown in Table 8.

     The relations between coal reactivity and several para-
meters are shown in Figs. 3 to 8. In these figures the reactivity of coal is measured by conversion. In the results, volatile
carbon % is selected as a more closely related parameter than
the common parameters, such as C%, H%, O%, H/C atomic ratio,
volatile matter, etc.

     Volatile carbon % is defined by the equation as follows.

Volatile carbon %

$$= C\% \text{ (d.a.f.)} - \frac{\text{Fixed carbon \%}}{\text{Volatile matter \% + Fixed carbon \%}} \times 100$$

     This parameter is derived from the following idea.
It is generally considered that the first step of coal hydroliquefaction is the thermal decomposition of C-C and C-O bonds,
etc. in coal structure. Thus, it is presumed that the volatile
matter in coal is closely related, as a parameter to coal reactivity (conversion). But, the amounts of oxygen containing
compounds, such as carbon dioxide, water, etc. in volatile matter
formed by the thermal decomposition of oxygen containing functional groups in coal, are large and vary greatly with the rank
of coal. Moreover, the functional groups are mostly attached to
the side chain of the basic aromatic units in the coal structure.
Thus, the volatile matter in coal is not generally considered to
be a better parameter representing coal reactivity.

Table 6.   Analytical data for coals used and experimental results

(Yamakawa et al)

Original coals	Ultimate analysis (d.a.f.)					Proximate analysis			Characteristics			Conv. (%)
	C	H	O	N	S	A	V M	F C	Ro	Inerts	V C	
Illinois	73.8	5.8	17.5	1.8	1.1	7.1	41.4	51.5			20.1	94.7
Kentucky NO.11	75.5	5.9	13.7	1.2	3.7	2.5	58.8	38.7			36.0	80.6
Griffin	75.8	4.0	18.5	1.6	0.2	13.2	51.3	35.5	0.33	39.0	36.1	38.5
Taiheiyo	77.0	6.0	15.4	1.3	0.3	14.7	43.7	41.7	0.57	6.6	33.4	64.2
Miike	82.2	6.3	7.7	1.3	2.5	14.3	27.6	58.1	0.79	9.0	15.4	91.5
Gross valley	83.2	4.8	9.9	1.7	0.4	8.9	39.5	51.5	0.83	60.4	26.9	34.8
Newdell	83.5	5.8	8.3	1.9	0.5	20.0	27.4	52.6	0.80	20.5	19.4	57.3
Hwaipei	85.2	5.0	8.3	1.3	0.3	2.7	46.0	51.4	0.99	24.2		45.9
Wallondilly	85.9	5.3	6.5	1.8	0.5	9.5	29.0	61.6		35.5	33.6	56.9
Yubari	86.3	6.2	5.3	1.9	0.3	11.1	20.9	68.0	1.11	44.8	19.0	95.2
Wollondilly	87.0	5.2	5.7	1.8	0.4	9.8	22.7	67.6	1.52	41.7	11.8	44.6
Weathered Balmer	88.4	4.4	5.7	1.2	0.3	6.2			1.46	32.2	14.2	23.8
Balmer	89.1	5.5	3.9	1.1	0.4							41.6
South Yakution 0–6	89.9	4.6	4.3	0.9	0.3	8.0	19.9	72.1	1.70	22.0	11.5	28.2
Smoky River	90.6	4.7	2.9	1.3	0.5	7.7	19.4	72.9	1.75	29.1	11.6	33.0
South Yakutian 0–1	91.7	4.7	2.2	1.1	0.4	8.9	18.4	72.8	1.74	21.9	11.8	32.1

* dry base

Table 7.   Analytical data for coals used and experimented results

(P.H. Given et al)

Original coals PSOC NO.	Ultimate analysis (d.a.f.)					Proximate analysis			Characteristics				Conv. (%)
	C	H	O	N	Sorg	A*	V M**	F C**	Ro	RM***	In-erts	V C	
87	72.0	5.2	21.7	0.6	0.5	10.6	53.6	46.4	0.30	72	18	25.6	95.9-97.1
99	72.5	5.2	20.9	0.8	0.6	27.3	67.1	32.9	0.31	90	9	39.6	78.8
151	78.3	5.8	14.3	1.2	0.4	5.7	45.4	54.6	0.40	81	9	23.7	72.1
187	80.5	5.6	12.2	1.0	0.8	7.5	39.0	61.0	0.64	90	9	19.5	88.6
185	81.8	5.8	9.5	1.1	1.7	16.1	42.4	57.6	0.55	93	7	24.2	88.2
105	81.9	5.6	11.0	0.4	1.1	13.4	37.6	62.4	0.73	83	15	19.5	80.8
68	82.0	5.7	10.3	1.5	0.6	6.0	41.3	58.7	0.64	83	16	23.3	86.4
70	82.4	4.7	10.7	0.8	0.4	29.4	40.4	59.6	0.77	97	2	22.8	90.7
160A	82.5	5.4	9.8	1.8	0.5	6.2	41.1	58.9	0.75	87	11	23.6	92.4
95	84.4	5.9	7.6	1.1	1.0	23.8	39.9	60.1	0.84	94	3	24.3	97.6
110	85.8	5.7	6.6	1.3	0.7	7.4	37.6	62.4	0.93	90	9	23.4	87.8

* Dry base   ** d.a.f. base   *** RM means reactive maurals' content in coal.

Table 8. Comparison of the experimental conditions

	Present study		Yamakawa's	P.H. Given's
Rank of coals	from lignite to bituminous (4 coals)	lignite (5 samples)	from lignite to bituminous (16 coals)	from lignite to bituminous (11 coals)
Solvents	creosote type	creosote oil	creosote oil	anthracene oil
Catalysts	Fe_2O_3 + S no catalyst	Fe_2O_3 + S	no catalyst	used
H_2 initial press (kg/cm^2)	60	80	30	14 (atm.)
Reaction temp. (°C)	450	430	420	385
Reaction time (hr.)	2.0	1.0	0.5	1.0
Reaction press. (kg/cm^2)	140 - 190	195 - 248	-	238 (atm.)

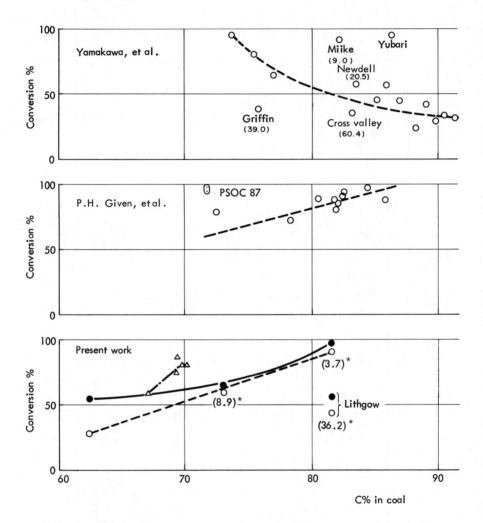

Figure 3. Relationship between conversion and carbon percentage in coal. The asterisks indicate that the figures in parentheses show the inert content in the coal. Symbols: (●), with catalyst; (○), no catalyst; (△), Morwell brown coal.

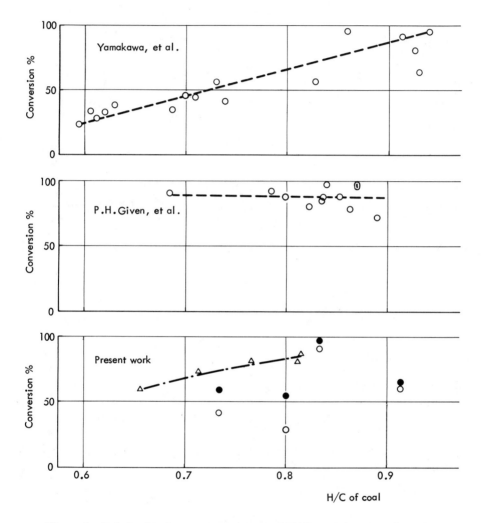

Figure 4.   Relationship between conversion and H/C of coal: (●), with catalyst; (○), no catalyst; (△), Morwell brown coal.

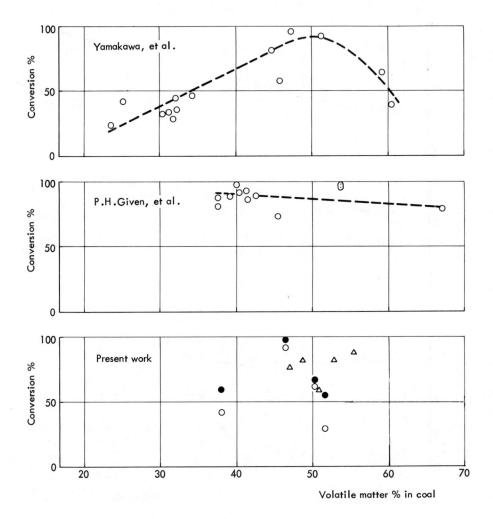

*Figure 5.   Relationship between conversion and volatile matter percent in coal:
(●), with catalyst; (○), no catalyst; (△), Morwell brown coal.*

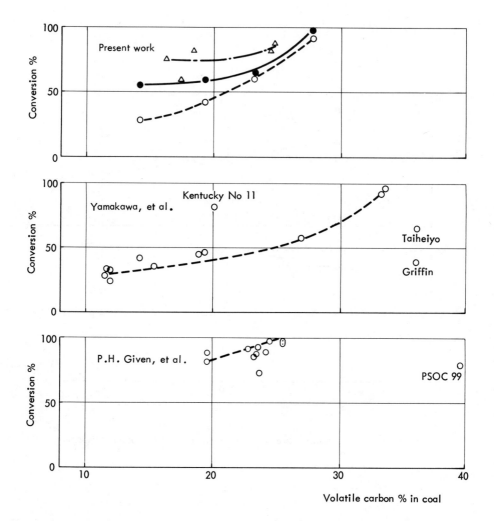

Figure 6. Relationship between conversion and volatile carbon percent in coal: (●), with catalyst; (○), no catalyst; (△), Morwell brown coal.

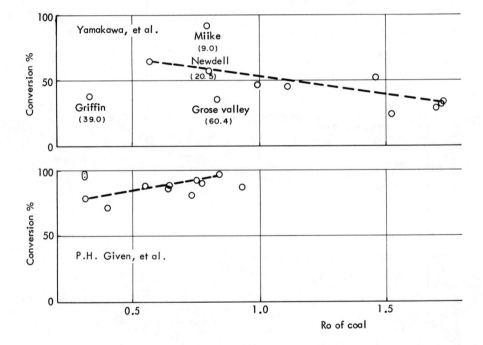

Figure 7.   Relationship between conversion and mean maximum reflectance

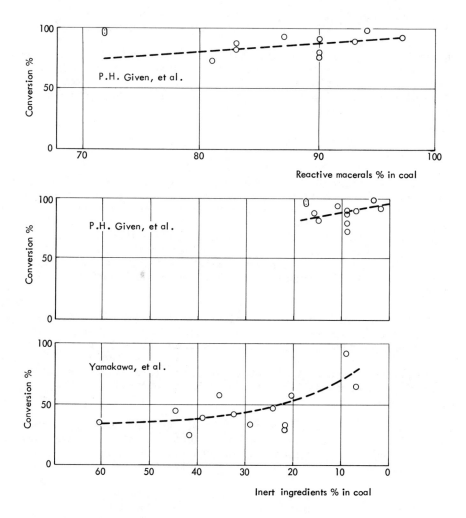

Figure 8. Relationship between conversion and petrographic components percent in coal

Based on the idea mentioned above, we should pay attention to the quantity of carbon content in the volatile matter in coal.

Discussion

It is well known that coal reactivity depends on the solvent, the conditions of hydroliquefaction, and the composition of the coal. Different extracting solvent results in different conversion, but it can be considered that the different conversion shows a similar tendency to coal reactivity. Thus, it is desirable that the parameter representing coal reactivity shows essentially the same tendency, despite the conditions of hydroliquefaction. Accordingly comparison of parameters was carried out, using some previously reported results (2, 3).

The following results were obtained as shown in the example below.

The relationship between conversion and C% in coal (d.a.f.) is shown in Fig.3. In this figure, the relatively close relationship between conversion and C% in coal is observed, but at the same time, it is found also that there are some exceptions in this relationship. The behaviour of abnormal coals could possibly be explained by the inert content in the coal at the same carbon level. That is, in Yamakawa's data, the inert content of Miike and Yubari coals are lower, while Griffin coal is higher.

Moreover, in coals of a similar carbon level, such as Miike coal, Newdell coal, and Grose valley coal, the reactivity of coal decreases greatly with the increasing inert content of coal.

The same result can be observed in our results. The inert content of Lithgow coal is fairly high as compared with the other coals used.

Furthermore, in P.H. Given's data, the lignite sample, PSOC 87 coal is very reactive though its inert content is higher, and deviates considerably from the general tendency. This seems to indicate that this coal was chemically treated.

From the data mentioned above, it is found that the consequences of this relationship depend on the conditions of liquefaction and coal quality used (Fig.3). Thus, C% in coal is not appreciably useful as a parameter. Similar consequences are found in the relationship between conversion and other parameters, such as H%, O% in coal.

The relationship between conversion and the H/C atomic ratio

of coal is shown in Fig.4. A fairly good relationship is found
in some restricted conditions of liquefaction, such as in
Yamakawa's, P.H. Given's and our data of Morwell brown coal, but
since there is no definite tendency in the several ranks of coal
we used, the characteristics of this relationship do not seem to
be general. Thus, the H/C of coal is not particularly useful as
a general parameter.

   The relationship between conversion and the volatile matter
% in coal is shown in Fig.5. According to Yamakawa's data,
conversion becomes higher with an increase in the volatile matter
content in coal in the range of 20% to 50%, but conversion
reaches a maximum of about 50%, and decreases after that. Roughly speaking, in P.H. Given's data, conversion decreases with the
increasing volatile matter % in coal in the range of 40% to 70%.
On the other hand, in our data, no clear relationship between
conversion and volatile matter% in coal can be found in the several ranks of coal and in the samples of similar rank levels.
The characteristics of this relationship are found to differ
greatly from one another in the conditions of liquefaction and
the coals used. Thus, the volatile matter % in coal is not
particularly useful as a parameter.

   The relationship between conversion and volatile carbon %
in coal is shown in Fig.6. As shown in this figure, conversion
of almost all coals in our research can be expressed exclusively
under the same experimental conditions. It was further found
that the effect of a catalyst was larger in coals of a lower
volatile carbon %. In Yamakawa's data, a fairly good relationship is found except for abnormal coals of high sulphur content
(Kentucky No.11) and of high inert content (Griffin), though the
behavior of Taiheiyo coal can not be explained. In addition, in
P.H. Given's data, a similar relationship, roughly speaking, is
found except for the abnormal coal (PSOC 99). In spite of the
differences of the liquefaction conditions and the coals used,
the characteristics of this relationship are almost the same
except for some abnormal coals. Therefore, it is safe to say
that coals of a higher volatile carbon % are more reactive than
those of lower volatile carbon %. Thus, volatile carbon % does
seem to be a better parameter to estimate coal reactivity. However, further study is necessary to clarify the validity of this
new parameter.

   The relationship between conversion and the mean maximum
reflectance is shown in Fig.7. In Yamakawa's data, a fairly
good relationship is found except for abnormal coals of high
inert content (Griffin, Grose valley) and low inert content
(Miike). A similar good relationship is also found in P.H.
Given's data. However, the characteristics of this relationship
are the reverse in both cases. Thus, it seems that the mean

maximum reflectance of coal is not useful as a parameter.

Recently the petrographic components' content in coal has been widely used as a new measure for finding the characteristics of coal which were treated only in an average manner until recently by the volatile matter % in coal or the carbon % in coal, etc. Thus, the relationship between the conversion and the petrographic components' % in coal is shown in Fig.8. In P.H. Given's data, a good relationship between the conversion and reactive macerals % in coal can be observed. Furthermore, a fairly good relationship between the conversion and inert ingredients % in coal can also be observed in both P.H. Given's and Yamakawa's data. And the characteristics of this relationship are essentially the same for the two different liquefaction conditions. Thus, it is concluded that the reactive macerals % or the inert ingredients % in coal is a better parameter to estimate coal reactivity.

As stated before, volatile carbon % is considered to be one of the most important parameters of hydroliquefaction. Also a fairly good linear relationship between the volatile carbon % in coal and low temperature tar yield from coal is found in Morwell brown coals, based on the data from the State Electricity Commission of Victoria (SECV) in Australia, as shown in Fig.9. Therefore, the low temperature tar yield is also estimated to be an important parameter. In addition, the color tone of brown coal (lithotypes) is shown in this figure. From this figure, it is observed that both volatile carbon % and low temperature tar yield are in a fairly good relation to the color tone of brown coal. Thus, as proposed by the Australian researchers, the color tone of brown coal is considered to be an important parameter.

Finally, we have made an effort to clarify the relation between the characteristics of coal and its reactivity. As coal is a complicated organic high molecular compound containing different kinds of inorganic ingredients, it seems difficult to clarify this relation briefly. However, the following parameters are considered to be more effective than the others, if attention is paid only to the organic ingredients of coal.

1) the parameter related to coalification and volatile matter content in coal which we proposed.

    Volatile Carbon %

2) the parameter related to the petrographic components in coal proposed by P.H. Given et al and Yamakawa et al.

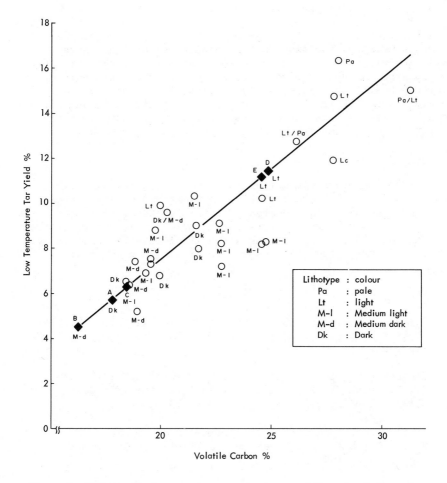

Figure 9. Relationship between low-temperature tar yield and volatile carbon content: (○), Australian researcher's data; (♦), present study data.

Reactive macerals % or Inert ingredients %

3) In the case of Australian brown coal, the parameter related to the color tone of coal proposed by the Australian researchers.

Lithotypes, etc.

The effectiveness of these parameters is considered to depend heavily on the liquefaction conditions and the characteristics of the coal which is used. The better parameters can possibly be derived from both the amounts of the petrographic components %, such as inerts' ingredients %, or reactive macerals % and their quality, such as H/C atomic ratio and so on. Consequently, it must be said that much further study is necessary to finally clarify the more comprehensive parameter.

Acknowledgement

Victorian brown coals (Yallourn, Morwell) used here and their data on coal characteristics were offered by the Herman Research Laboratory of the State Electricity Commission of Victoria, to whom the authors wish to express their appreciation. In addition, they wish to thank Nissho-Iwai Co. Ltd. who kindly acted as intermediary with respect to Victorian brown coals studied.

References

1. Dryden, I.G.C., Ed. "Chemistry of Coal Utilization, Supplementary Volume"; John Wiley and Sons; New York·London, 1962; P.237-244.

2. Yamakawa, T.; Imura, K.; Ouchi, K.; Tsukada, K.; Morotomi, H.; Shimura, K.; Miyazu, T.; "Preprints, the 13th Coal Science Conference". Japan, October, 1976, P.40.

3. Given, P.H.; Cronaner, D.C.; Spackman, W.; Lovell, H.L.; Davis, A.; Biswas, B. Fuel, 1975, 54, No.1, 40; OCR R&D Report No.61/Int. 9.

RECEIVED March 28, 1980.

Studies on Noncatalytic Liquefaction of Western Canadian Coals

B. IGNASIAK, D. CARSON, A. J. SZLADOW[1], and N. BERKOWITZ[2]

Alberta Research Council, 11315—87th Avenue, Edmonton, Alberta, Canada T6G 2C2

Though still only very incompletely explored and subject to major revisions, Canada's coal resources are so extensive as to place this country among the most richly coal-endowed nations (1,2,3). Recent appraisals (Table I) set ultimate in-place resources in >2½ ft thick seams under less than 2500 ft of cover at some 518 billion tons; and preliminary estimates from deeper testhole logs suggest that similar, if not even larger, tonnages may lie in coal occurrences at depths between 2500 and 4500 feet.

But there are wide regional disparities with respect to distribution and coal type.

Except for a relatively small (<500 million ton) lignite deposit in northern Ontario's James Bay area, the Central region (i.e., Quebec, Ontario and Manitoba), which accommodates some 70% of the country's population and the greater part of its industry, is devoid of coal; and the Maritime Provinces (principally Nova Scotia) contain less than 1% of Canada's total coal - mostly Carboniferous hvb coal which closely resembles its Eastern US counterparts.

The great bulk of Canadian coal is concentrated in the three Western provinces (Saskatchewan, Alberta and British Columbia). In this region, it is of Cretaceous and/or Tertiary age, with rank generally increasing in a westerly direction toward the Rocky Mountains. Although contained in different geological formations (which, in Alberta form three partially overlapping coal zones), the lignites of Saskatchewan thus give way to sub-bituminous coals in the Alberta Plains, and the latter successively to hvb, mvb and lvb coals in the Mountain regions further west.

Table I summarizes latest available data respecting reserves of these classes of coal.

[1] Current address: CANMET, Federal Department of Energy, Mines and Resources, Ottawa, Canada.
[2] Current address: University of Alberta, Department of Mineral Engineering, Edmonton, Alberta, Canada.

0-8412-0587-6/80/47-139-097$05.00/0

As matters stand, the low-rank coals of Western Canada (as
well as some hvb coal in the Maritimes) are now being increasingly used for generation of electric energy, and metallurgical (mvb
and lvb) coals are being primarily produced for export (notably
to Japan, though other markets are being developed in Korea,
South America and Western Europe). But the large reserves of
near-surface subbituminous coals and lignites are also being
looked upon as future sources of synthetic fuel gases and liquid
hydrocarbons that would augment production of synthetic crude
oils from, e.g., Northern Alberta's oil sands (4).

A notable feature of the Western Canadian coals is their low
sulphur content (usually <0.5%) which tends, however, to be
partly offset by higher mineral matter contents than are associated with the Eastern coals. As well, bituminous coals in the
mountain belts are typically deficient in vitrinite, which often
represents less than 50% of the coal "substance" and only occa-
sionally reaches 70-75%, but this is compensated by the fact that
their micrinites and semifusinites tend to be "reactive" constituents when the coals are carbonized. Notwithstanding their low
fluidity (rarely >1000 dd/min), Western mvb coals therefore make
excellent metallurgical cokes when carbonized in suitably propor-
tioned blends.

But, perhaps reflecting their unique petrographic make-up
as much as a more basic chemistry which may set them apart from
their Carboniferous equivalents, the Western coals also tend to
respond differently to, e.g., oxidation and the action of solvents on them. Air-oxidation at 150°C, instead of developing
acid oxygen functions, incorporates much of the chemisorbed oxygen in carbonyl groups; and solubility in $CHCl_3$ (after shock-
heating to $\sim$400°C) is substantially smaller than the FSI would
lead one to expect from correlations for Carboniferous coals (5).

These, and other, behaviour differences have prompted initiation of several exploratory studies in order to assess the
response of selected Western coals to liquefaction procedures
and identify the parameters that affect this response. This
paper summarizes some of the more important observations recorded
in the course of that work.

1. Liquefaction (Solubilization) by Interaction with H-Donors

To test solubilization, $\sim$5 gm samples were reacted with 10-
15 gm тetralin at 390±°C and autogenic pressures in helium-
purged, sealed Pyrex capsules. (To counterbalance the pressures
generated in them, the capsules were inserted into a stainless
steel bomb charged with $\sim$30 ml тetralin.) Reactions were carried
to completion over 4 hrs, after which the capsules were cooled to
room temperature and opened in a manner that permitted quantitative analysis of all reaction products.

Residual Tetralin was then removed by heating the solvolyzed samples at 70-80°C in vacuo (∿0.05 mm Hg) to constant weight, and yields of non-volatile products and their solubilities in pyridine and in benzene were determined.

The solubilities thus recorded for 13 Western (Cretaceous) coals (with 69.6-91.5% carbon, daf) and 8 Carboniferous coals (80.6-90.9% C, daf) are shown in Figure 1, and indicate that

(a) the pyridine-solubilities of reacted Carboniferous subbituminous and bituminous coals are significantly higher than those of corresponding Cretaceous coals, and

(b) strongly caking Carboniferous coals (with 86-88% C, daf) tend to generate substantially more benzene-soluble matter than their Cretaceous counterparts under the conditions of these experiments.

What is, however, still unclear is whether these effects arise solely from different chemical compositions (and molecular configurations) or are also, at least in part, a consequence of the Cretaceous coals generally containing almost twice as much mineral matter as the Carboniferous samples.

2. The Role of Ether-Linkages in Solubilization of Low-Rank Carboniferous Coals by H-Donors

Formation of asphaltenes during solubilization of low-rank bituminous coals has been attributed to cleavage of open ether-bridges (6). But while the presence of such configurations in high- and medium-rank bituminous coals is well established (7), their existence in less mature coals remains to be demonstrated. From reactions of low-rank bituminous coals with sodium in liquid ammonia or potassium in tetrahydrofuran, it has, in fact, been concluded that open ether-bonds are absent (8) or only present in negligible concentrations (9).

The failure to detect open ether-linkages by treatment with Na/liq. NH_3 could conceivably be due to formation of non-cleavable phenoxides (10). We note, in this connection, that low-rank coals, which contain much "unreactive" oxygen, are also characterized by relatively high concentrations of hydroxyl groups, and some "unreactive" oxygen could therefore be quite reasonably associated with phenoxy phenol configurations. However, whereas phenoxy phenols would be expected to resist cleavage by hydrogen-donors, low-rank coals are, as a rule, most easily solubilized by them; and this seeming inconsistency has prompted us to reexamine the behaviour of oxygen-linkages during interaction with H-donors.

The reactions were carried out under the same conditions as solubilization (see sec. 1), except that a constant 2:1 donor: substrate (molar) ratio was used; and for comparative purposes, all runs with Tetralin were repeated with 1,2,3,4-tetrahydro-

Table I. Canada's ultimate in-place coal resources
(Energy, Mines & Resources, Canada
Report EP 77-5, 1976 Assessment)

Region	In-place, billion tons	Principal coal type
Maritime Provinces Nova Scotia New Brunswick	1.7	hvb (Carb.)
Western Region Saskatchewan	38.8	lignite (Tert.)
Alberta* - Plains	360.0	subbit. (U.Cret. & Tert.)
Foothills	10.0	lvb, mvb, hvb, (Cret.)
Mountains	30.0	lvb, mvb, hvb, (Cret.)
British Columbia	77.8	mvb, lvb (Cret.) subbit. (Tert.)
Canada Total	518.3	

*After Energy Resources Conservation Board, Province of Alberta, Report 77-31, December 1976.

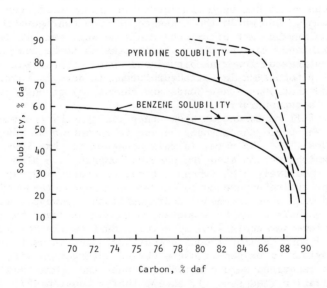

Figure 1.   Solubilities after treatment with tetrahydronaphthalene at 390°C for 4 hr: (———), cretaceous coals; (– – –), carboniferous coals.

quinoline as the donor. The results obtained with different ethers are summarized in Table II.

Detailed discussion of these findings will be presented elsewhere. Here we only wish to point out that responses to a hydrogen donor tend to be critically affected by minor structural differences between the compounds. Thus, while diphenyl ether remains substantially unaffected by the donor, its hydroxy-derivatives (phenoxy phenols) often display fairly high reactivity. Taken in conjunction with the failure of low-rank coals (7) and phenoxy phenols (10) to suffer reductive cleavage when treated with sodium in liquid ammonia, this lends some support for the existence of phenoxy phenol entities in low rank coals.

Other observations, however, indicate that this notion will require more direct evidence before it can be accepted.

The inertness of phenols and phenoxy phenols toward Na/liq. NH_3 can be attributed to the fact that phenols are powerful proton-donors in this system, and resistance of the resultant anions toward reduction is believed to result from stabilization by resonance (10). While alkylation of low-rank coals before treatment with Na/liq. NH_3 therefore offers means for establishing the presence of phenoxy phenol ethers in them, an alternative is afforded by the observation that some phenols can be reduced by concentrated solutions of lithium (11). If this latter reaction also reduces phenoxy phenols in coal, a second treatment should then cause ether-cleavage.

We found, however, that even highly concentrated lithium (9M) or sodium (3M) solutions did not reduce coal in a manner that increased its hydroxyl content; and in parallel tests, 100% unreacted p-phenoxy phenol was always recovered from the lithium solutions.

The failure to cleave p-phenoxy phenol by reduction and subsequent scission of the ether-bond led us to examine the possibility of splitting the C-O bond in the alkylated molecule (12). Attempts to alkylate p-phenoxy phenol with C_2H_5Br after treatment with lithium in liquid ammonia were unsuccessful; but ethylation in Na/liq. NH_3 yielded nearly 50% of the ethylated product, and ethylation in K/liq. NH_3 led to quantitative conversion of the substrate - with 40% of the reaction product recovered as p-diethoxybenzene. Formation of p-diethoxybenzene on treatment of p-PhO-PhOH with potassium followed by alkylation with C_2H_5Br results from cleavage of p-PhO-$PhOC_2H_5$ (formed in early stages of alkylation) in presence of residual amounts of potassium. Protonation (CH_3OH) of potassium salts formed by reacting p-PhO-PhOH with K/liq. NH_3 did not form p-dihydroxybenzene.

Treatment of low-rank coal (or of a vitrinite fraction from such coal) with variously concentrated solutions of potassium in liquid ammonia did <u>not</u> cause an increased -OH content in the reacted material. Nor was the hydroxyl content affected by such treatment after prior exhaustive methylation of the coal with dimethyl sulphate and K_2CO_3 in acetone (13). On the other hand,

TABLE II. Reactions of Ethers with

Ether	Structure	% of Ether Conversion	Hydrogen Consumption Moles/Mole	Recovery of Tetrahydronaphthalene + Naphth. [%]	Reaction Products Identification	Yield [Molar %]
dibenzyl ether	⬡-CH₂-O-CH₂-⬡	100	0.05	~100	toluene benzene benzaldehyde 1 methylindane	50 29 16 undetermined
p(benzyloxy)phenol	⬡-CH₂-O-⬡-OH	-	0.8	~100	toluene	70
benzyl 1-naphthyl ether	⬡-CH₂-O-⬡⬡	-	0.75	~100	toluene bibenzyl 1-naphthol 2-benzylnaphthalene 1-methylindane(1) 3 unidentified	73 7 74 undetermined undetermined -
diphenyl ether	⬡-O-⬡	2	0.0	~100	no GC detectable prodts.	-
p-phenoxyphenol	⬡-O-⬡-OH	7	0.04	88	phenol	7
m-phenoxyphenol	⬡-O-⬡ OH	0	0.0	~100	none	-
m-diphenoxybenzene	⬡-O-⬡-O-⬡	0	0.0	100	none	-
p-phehoxybiphenyl	⬡-O-⬡-⬡	0	0.0	100	none	-
furan	⬠	0	0.0	100	none	-
2,3-benzofuran	⬡⬠	0	0.0	100	none	-
dibenzofuran	⬡⬠⬡	0	0.0	~100	1 unidentified	trace
tetrahydrofuran	⬠	0	0.0	~100	1 unidentified	trace

1. probably originating from donor
2. % by weight
3. elemental analysis of the high molecular weight product unknown
4. elemental analysis: 82.3%C; 7.0%H; 4.2%N; 0.0%S; 0.0% ash; 6.5%0 (by diff.)
5. elemental analysis: 79.0%C; 6.2%H; 1.8%N; 0.0%S; 0.0%ash; 13.0%0 (by diff.)

THN and Tetrahydroquinoline at 385°C

% of Ether Conversion	Hydrogen Consumption Moles/Mole	Recovery of Tetrahydro-quinoline + Quinoline [%]	Identification	Yield [Molar %]
			1,2,3,4 Tetrahydroquinoline	
			Reaction Products	
100	~1	85	toluene	51
			water	10
			3 methylpyridine(1)	undetermined
			ethylbenzene(1)	undetermined
			high mol.w.product(3)	undetermined
-	-	40	toluene	72
			high mol.w.product(3)	undetermined
0	0.0	~100	no GC detectable prodts.	-
50	-	62	phenol	43
			o-toluidine(1)	undetermined
			high mol.w.product(4)	90(2)
24	-	73	high mol.w.product(5)	90(2)
0	0.0	100	none	-
0	0.0	100	none	-
0	0.0	100	none	-
0	0.0	100	none	-
0	0.0	~100	2 unidentified	traces
0	0.0	100	none	-

alkali metal reduction of methylated, high molecular weight,
complex phenoxy phenol type compounds always resulted in ether
cleavage (11).

It appears to us therefore that cleavage of ether-bonds
contributes little to solubilization (and consequent reductions
of the molecular weight) unless the coal contains an appreciable
proportion of open oxygen linkages in the form of dialkyl ethers.
And since there are indications that such structures are generally absent (14), one might tentatively conclude that molecular
weight reductions during solubilization by H-donors accrue primarily from C-C bond scission or from structural realignments
associated with elimination of oxygen. It should be possible to
test this by measuring molecular weight distributions in H-donor
liquefied and non-destructively solubilized coal products (see
sec. 3).

3. Non-Destructive Solubilization of Low-Rank
    Bituminous Coal (by non-reductive alkylation)

    Present methods for solubilizing coal (including reductive
alkylation in tetrahydrofuran (15) or liquid ammonia (8)) entail
cleavage of oxygen ethers, scission of C-C bonds in certain
polyaryl-substituted ethylenes and, in the case of reactions in
tetrahydrofuran, extensive elimination of hetero-atoms (16).

    We therefore draw attention to a novel technique which
allows solubilization of coal without rupture of covalent bonds.
This utilizes the fact that the acidity of low-rank coals, which
is largely due to their high -OH contents, can be enhanced by
proper choice of a medium.

    We selected liquid ammonia because of its pronounced solu-
bilizing characteristics and powerful ionizing properties. At
-33°C and atmospheric pressure, the pK_a-value for auto-ioniza-
tion of liquid ammonia [$2NH_3 = NH_2^{\ominus} + NH_4^{\oplus}$] is 34; and since
the equivalent value for water is only 14, many substances (with
pK_a-values between 14 and 34) which are neutral in water should
be capable of splitting off protons in liquid ammonia. Acidic
properties in liquid ammonia can be further enhanced by increas-
ing the concentration of $NH_2^{\ominus}$ at the expense of protonic $NH_4^{\oplus}$;
and this can be achieved by adding potassium and/or sodium
amides which will then also form the respective coal "salts".

    To test this approach, ∿5 g samples -300 mesh Tyler, of a
low-rank vitrinite, were stirred for 6 hrs in liquid ammonia
(150 ml; -33°C) containing ∿5 gms of potassium amide and ∿5 g
of sodium amide. (The amides were formed in the medium, before
introducing the coal, by action of anhydrous ferric oxide (1 g)
or ferric chloride (1.5 g) on alkali metals.) Thereafter, 100
ml of anhydrous ethyl ether was added, the suspended coal material ethylated with C_2H_5Br (32 ml), and the reaction mixture
stirred until all ammonia and ether had evaporated. Following

acidification of the residue with 10% HCl, the product was thoroughly washed with distilled water, dried at 70-80°C in vacuo (0.05 mm Hg) and analysed. Table III summarizes the results of three consecutive alkylations, with each datum being the average of four independent test runs. The initial ethylation introduced 7-8 ethyl groups/100 C atoms into the coal, and the results of the second and third ethylations indicate that essentially only -OH groups were ethylated at this stage. Overall, however, over 50% of ethyl groups introduced into coal were not linked to hydroxyl functions and it is therefore tentatively concluded that low-rank coals contain a significant number of acidic carbon atoms. It is known that the acidity of phenyl-substituted alkanes in which at least two phenyl groups are attached to the same carbon atom is sufficiently high to allow proton abstraction in liquid ammonia (11).

The most interesting outcome of this work is the observation that low-rank vitrinites can be rendered substantially soluble in chloroform and pyridine by alkylating coal salts formed in a non-reducing medium and under conditions that appear to preclude cleavage of covalent bonds.

In coals alkylated in this manner, the number of acidic sites is substantially reduced, and acid-base associations are virtually precluded. Extracts from alkylated coals should, therefore, be amenable to GPC fractionation. Such fractionation, conducted on Bio Beads S-X1 and S-X2, results in separation by molecular weight and indicates that both benzene and chloroform extracts contain substantial amounts of high ($\sim$6000) and fairly low (560-640) molecular weight fractions (Figure 2). While the extract yields from non-reductively ethylated vitrinite increase in the order benzene extr. → chloroform extr. → pyrid. extr., the molecular weights determined by VPO in pyridine, decrease in this order.

Figure 3 shows GPC fractionation of the benzene extract of a vitrinite which, before a single non-reductive ethylation, was treated with tetrahydronaphthalene at 390°C. Although hydrogenation reduced the -OH contents (from 4.9 to 1.7%), non-reductive alkylation increased the benzene-solubility of the solvolyzed material from 53% to 80.2%. It appears that this effect is due to ethylation of acidic atoms.

Analysis of the results presented in Figures 2 and 4 appears to indicate that hydrogenation of the vitrinite is also accompanied by polymerization (see pyridine extract, Figure 4). If this can be confirmed, it would be worth investigating whether it involves specific fractions in the original vitrinite or has a random character. Solvolysis of different molecular weight fractions of a non-reductively alkylated vitrinite (or coal) may furnish some insight.

TABLE III.

Non-reductive ethylation of a
Carboniferous vitrinite* in liquid ammonia

	Ethyl groups intro. per 100 mg Cat.	Chloroform solubility	Pyridine solubility	Ultimate analysis (daf)			No. of hydroxyl groups per 100mg Cat.
				%C	%H	%N	
I Ethylation	7-8	25	45-50	81.0	6.3	1.6	2.8
II Ethylation	8-9	45-50	55-60	81.3	6.8	1.8	2.0
III Ethylation	9-10	55-60	60-65	81.1	7.0	1.5	1.1

*Solubilities of untreated vitrinite: $CHCl_3$ - 1%; C_5NH_5 - 13%;

Elemental analysis of untreated vitrinite (daf): 80.8%C; 5.2% H; 1.5% N; 0.9% S; 5.0% O_{OH}

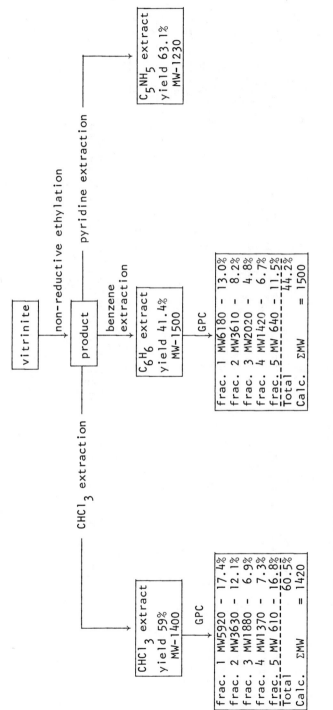

Figure 2. *Solubilities and molecular weights of a vitrinite (80.8% C, daf) after nonreductive ethylation (yields calculated on ethylated product (daf))*

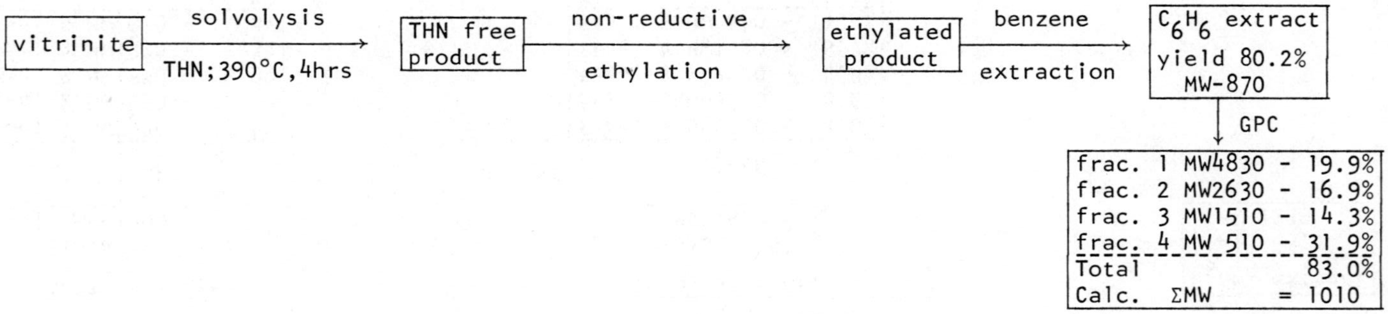

Figure 3. Solubilities and molecular weights of a vitrinite (80.8% C, daf) subjected to solvo-lysis in THN prior to nonreductive ethylation (yields calculated on ethylated product (daf))

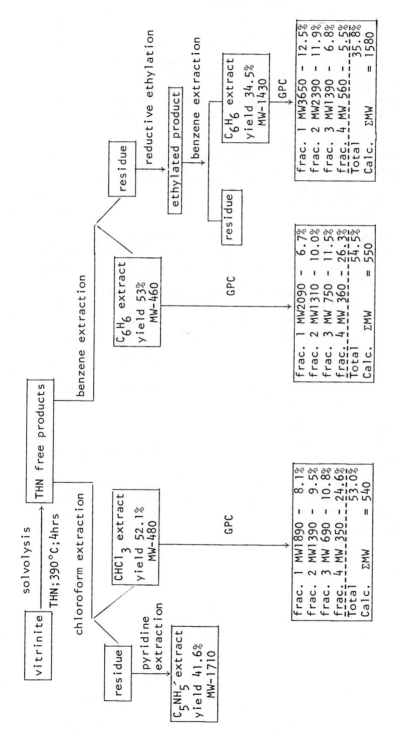

Figure 4.   Solubilities and molecular weights of a vitrinite subjected to solvolysis in THN (yields calculated on THN free product (daf))

Literature Cited

1. Ruedisili, L. C. and Firebauch, M. W., Perspectives on Energy, New York-Oxford University Press - London - Toronto, 1975.
2. Reserves of Coal - Province of Alberta, Energy Resources Conservation Board, Calgary, Alberta, Canada, T2P 0T4 Dec. 1976.
3. 1976 Assessment of Canada's Coal Resources and Reserves; Energy, Mines & Resources, Canada, Report EP 77-5.
4. Berkowitz, N. and Speight, J. G., Fuel, 1975, 57, 138.
5. Berkowitz, N., Fryer, J. F., Ignasiak, B. S. and Szladow, A. J., Fuel, 1974, 53, 141.
6. Takegami, Y., Kajiyama, S. and Yokokawa, C., Fuel, 1963, 42, 291.
   Whitehurst, D. D., Farcasiu, M., Mitchell, T. O. and Dickert, J. J., The Nature and Origin of Asphaltenes in Processed Coals. EPRI AF-480 (Research Project 410-1), July 1977.
7. Lazarov, L. and Angelova,G., Fuel, 1968, 47, 333.
8. Ignasiak, B. S. and Gawlak, M., Fuel, 1977, 56, 216.
9. Wachowska, H. and Pawlak, W., Fuel, 1977, 56, 422.
10. Tomita, M., Fujita, E. and Abe, T. J. Pharm. Soc. Japan 1952, 72, 387;
    Tomita, M., Fujita, E. and Niva, H., ibid. 1952, 72, 206.
11. Harvey, R. G., Synthesis 1970, No. 4, 161;
    Smith, H. Organic Reactions in Liquid Ammonia, Inter-science Publishers, 1963.
12. Tomita, M., Fujita, E. and Murai, F. J. Pharm. Soc. Japan 1951, 71, 226 and 1035;
    Tomita, M., Inibushi, Y. and Niva, H., ibid. 1952, 72, 211.
13. Briggs, G. C. and Lawson, G. J., Fuel, 1970, 49, 39.
14. Pugmire, R. J., personal communication.
15. Sternberg, H. W., Delle Donne, C. L., Pantages, P., Moroni, E. C. and Markby, R. E., Fuel, 1971, 50, 432.
16. Ignasiak, B. S., Fryer, J. F. and Jadernik, P., Fuel, 1978, 57, 578.

RECEIVED March 28, 1980.

Reactivity of British Coals in Solvent Extraction

J. W. CLARKE, G. M. KIMBER, T. D. RANTELL, and D. E. SHIPLEY

National Coal Board, Coal Research Establishment, Stoke Orchard, Cheltenham, Gloucestershire, England

By 1980 the rate of recovery of light crude oil from the North Sea oil fields will exceed the total demand for crude oil in the United Kingdom. Heavy crude oils from the Middle East will be used for balancing the refineries to produce the required range of petroleum products. However, by the late 1980's the diminishing supply of indigenous oil and general world shortage will result in a serious shortfall in supply. Thus it will be necessary to exploit other more abundant resources of hydrocarbons. In the United Kingdom there are large reserves of coal which could satisfy demand for at least two hundred and fifty years. The National Coal Board currently mines approximately 120 million tons of coal per annum of which 65% is used for generating electricity ($\underline{1}$). (Combustion of the low sulphur British coals does not result in excessive atmospheric pollution.) To satisfy the increased demand for coal the National Coal Board has undertaken an investment programme which includes the development of a mining complex at Selby in Yorkshire which, it is estimated, will produce in excess of 10 million tons when full production is reached in 1988.

At the Coal Research Establishment of the National Coal Board, methods for the liquefaction of coals to produce transport fuels, feedstocks for the chemical industry and high purity carbons suitable for electrode manufacture are being developed. A schematic diagram of the liquid solvent extraction process is illustrated in Figure 1. Where the production of liquid hydrocarbons is the main objective an hydrogenated donor process solvent is used, whereas in the production of needle coke this is not necessary and a coal derived high boiling aromatic solvent may be used (e.g. anthracene oil). An essential economic requirement of the process is that a high extraction yield of the coal is obtained and this will depend upon the coal used and the digestion conditions.

The properties of the coals mined in the United Kingdom vary from the high carbon anthracites to the lower rank non-

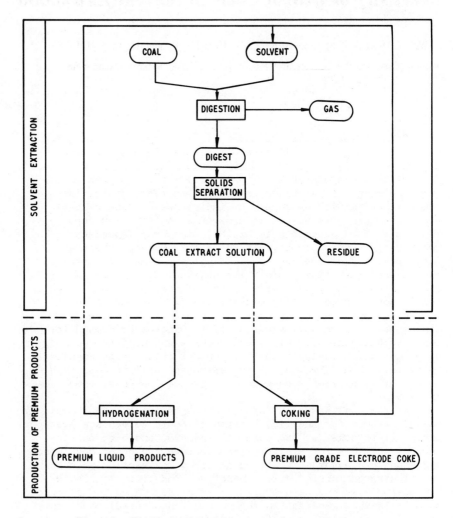

Figure 1.   The National Coal Board solvent extraction process

caking coals, but with virtually no deposits of brown coals or lignites. These variations in composition can be conveniently illustrated on Seyler's coal chart (Figure 2) (2). In general each mining area produces coal of a characteristic type, for example, the South Wales coal fields produce anthracites and prime coking coals; while the large mining areas of Yorkshire, Derbyshire and Nottinghamshire produce lower rank coals and it is here that the substantial reserves of coal are found.

Extensive studies into the solvent extraction of coals have revealed the mechanism controlling digestion and produced kinetic data on the reactions (3-11). In digestion, a coal-solvent slurry is heat-treated at temperatures in the range 400-440°C during which most of the coal depolymerises and dissolves. The heat treatment causes some bonds in the molecular structure of the coal to rupture producing radicals which are stabilised by hydrogen transfer from the solvent. In the absence of hydrogen the radicals combine to form an insoluble high molecular weight product. This is a simplified model and more detailed studies of digestion are available (8, 9). Kinetic studies have mainly supported this model but the interpretation of the reaction orders and activation energies of the reactions appears complex (10, 11). In a study of the digestion of a low rank British coal in anthracene oil it was found that the initial decomposition of the coal was of first order with an apparent activation energy of 40 kJ mole^{-1}. This was followed during further heat-treatment of the digest by a second order polymerisation with an activation energy of 120 kJ mole^{-1}. For a solution prepared from a high rank coal the polymerisation was again second order but with an activation energy of 190 kJ mole^{-1} (12).

Most studies on the mechanism and kinetics of solvent extraction have necessarily used only a limited selection of coals. In a commercial environment where coals with widely varying properties are available, it is necessary to develop a generalised system of grading the coals with respect to their suitability for liquefaction.

To classify coals for the production of metallurgical cokes the National Coal Board adopted a system of ranking based upon the Gray King coke type and volatile matter (Figure 2) (13). Unfortunately, this method of ranking coals was found unsuitable for solvent extraction (14). In this work consideration has been given to methods of ranking coals based upon reactivity during liquefaction.

Technique

Samples of digest were prepared using either the maxibomb digester shown in Figure 3 or a smaller digestion vessel (minibomb) described elsewhere (12). The maxibomb consisted of a length of 25 mm o.d. stainless steel tube closed at one end

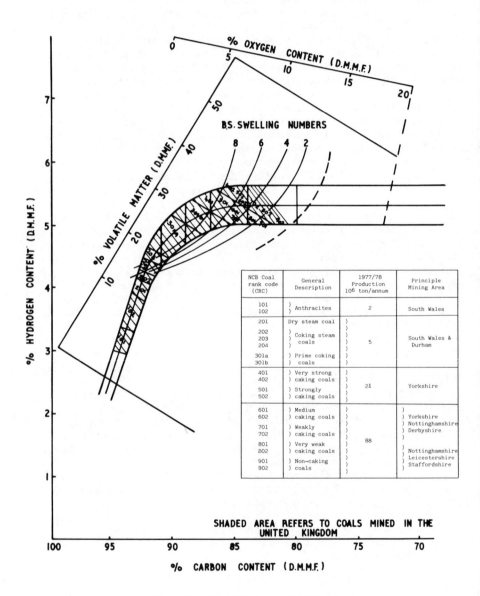

Figure 2.   Classification of British coals

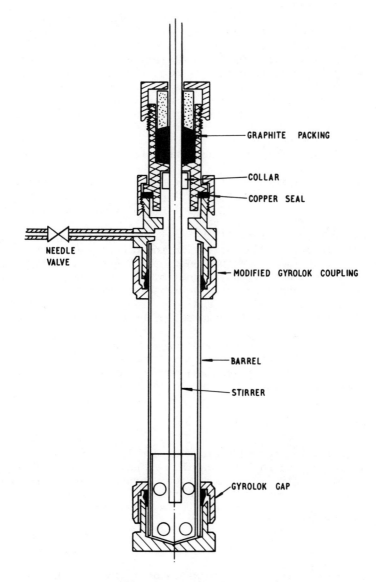

GRAPHITE PACKING

COLLAR

COPPER SEAL

NEEDLE VALVE

MODIFIED GYROLOK COUPLING

BARREL

STIRRER

GYROLOK GAP

Figure 3.   Maxibomb digester

with a 'Gyrolok' compression coupling. The top of the tube was
closed with a modified coupling containing a packed gland and
stirrer. Approximately 20 g of a coal-solvent slurry were
accurately weighed into the barrel of the reactor. The complete
bomb was immersed in a fluidised sandbath controlled at the
required operating temperature. After a given residence time
the bomb was removed from the sandbath and water quenched. The
volume of gas produced during digestion was measured.

For short residence times it was necessary to compensate
for the heating up period. Heating curves were constructed at
each digestion temperature by using a modified digestion vessel
into which a thermocouple had been sealed. From these curves
and a knowledge of the activation energies it was possible to
calculate the time interval to be allowed (seven minutes for
maxibomb) after immersion of the bomb in the sandbath before
digestion was effectively started (12). This method of
compensating for the heating up period introduces a small error
in the short residence time tests.

The extraction yield of the coal was determined from the
concentration of residual solids in the digest and the yield
of gas. The solids concentration of the digest was determined
from the measurement of the solubility of the digest in
quinoline. The addition of quinoline does not precipitate the
dissolved coal in the digest and thus the yield of quinoline
insolubles can be directly equated to the concentration of
solids in the digest (12).

In some tests the samples of digest were filtered at
constant temperature and pressure through a heated pressure
filter. The yield of filter cake was measured and the solubility
in quinoline determined in order to calculate the extraction
yield.

Materials

Data relating to the coals used are listed in Table I.
The coals were crushed 80% less than 75 μm. The solvents used
were anthracene oil (ex British Steel Corporation), hydrogenated
process solvent (produced in a continuous coal extract
hydrogenation plant) and several pure organic compounds
(ex Koch-light).

Results

The results from the extraction of coals in anthracene
oil and phenanthrene are given in Tables III to VII, while
Table II gives extraction yields of Annesley coal when digested
in a range of solvents including a selection of organic
compounds. The influence of digestion conditions upon extraction
is shown in Figures 4 to 6.

Table I

Analysis of Coals

Coal	NCB Coal rank code (CRC)	Proximate analysis (ar)			Ultimate analysis % (dmmf)				
		% Moisture	% Ash	% Volatile matter	C	H	O	N	S (total)
Cynheidre beans	101	1.1	7.4	5.4	93.7	3.1	1.2	1.2	1.0
Penalta	202	0.5	7.3	14.2	91.3	4.3	2.0	1.5	0.9
Garw	204	0.5	9.4	18.1	91.6	4.4	1.3	1.6	0.9
Tilmanstone	204	0.5	7.3	17.7	91.7	4.3	1.8	1.4	0.6
Tymawr	204	0.5	5.8	18.8	91.2	4.5	1.9	1.6	0.8
Beynon	301a	1.0	8.1	24.1	89.4	4.9	3.2	1.4	1.0
Windsor	301a	1.2	7.4	26.1	89.7	4.8	1.5	3.3	1.5
Marine	301a	1.3	10.2	25.1	90.4	5.0	1.6	2.7	0.8
Oakdale	301a	1.3	6.7	20.9	91.2	4.7	1.6	1.6	0.9
Cwm (a)	301a	0.5	2,3	21.3	91.0	4.8	1.9	1.5	0.8
(b)	301a	0.5	6.7	21.6	90.8	4.7	2.6	1.5	0.4
(c)	301a	0.8	2.7	21.7	90.3	4.9	2.3	1.6	0.9
(d)	301a	0.8	4.2	22.9	90.6	4.9	2.3	1.6	0.6
Bedwas	301b	0.9	10.7	28.5	88.8	5.1	1.5	3.8	1.7
Herrington	401	1.0	8.2	32.8	87.2	5.2	4.8	1.9	0.5
Bersham	502	1.5	4.5	39.2	85.7	5.2	6.0	1.7	0.9
Wearmouth	502	3.4	5.4	36.0	86.0	5.2	5.8	1.8	0.9
Hawthorn	502	2.3	7.3	35.5	86.8	5.4	5.3	1.8	1.5
Manton	502	2.3	3.7	37.0	85.7	4.7	6.1	1.4	2.5
Barrow	502	1.1	6.6	36.3	86.6	5.4	5.1	1.8	1.9
Brodsworth (a)	602	2.6	4.3	38.1	84.5	5.3	7.1	1.9	1.8
Swallowood	602	3.8	5.7	36.9	85.4	5.6	6.1	1.8	1.3
Brodsworth (b)	702	2.0	14.2	38.9	85.1	5.3	6.3	1.9	1.8
Annesley	702	1.8	6.3	37.9	84.5	5.4	8.0	1.9	0.7
Rufford (a)	702	5.4	5.6	38.8	84.1	5.3	7.7	2.0	2.0
Rufford (b)	702	5.3	5.9	38.7	84.3	5.2	7.6	1.9	1.3
Blidworth	702	6.9	8.1	40.4	83.6	5.0	8.3	1.8	1.6
Markham	702	5.5	5.0	39.2	–	–	–	–	–
Whitwell	902	3.1	5.7	38.6	83.2	5.0	8.7	2.0	0.9
Bestwood high	902	4.2	6.2	39.0	–	–	–	–	–
Linby (fresh)	802	6.9	7.5	34.3	84.6	5.5	7.3	1.9	0.8
Linby (aged)	902	3.1	7.5	38.0	82.5	5.3	9.5	1.9	0.8

Table II

Solvent Power

Coal: Annesley
Digestion: 400°C, 60 minute residence time
Coal:Solvent ratio 1:4

Solvent	Total extraction yield[1] (%)
Diphenyl	13
Naphthalene	14
Anthracene	32
1-Methylnaphthalene	48
Dibenzofuran	51
2-Methylnaphthalene	51
Phenanthrene	55
Fluorene	66
Dibenzothiophene	66
Anthracene oil	71
9,10-Dihydroanthracene	77
Pyrene	83
Acenaphthene	85
Tetralin	86
Hydrogenated anthracene oil	89
Indoline	95
1,2,3,4-Tetrahydroquinoline	95

Note 1 d.a.f. basis.

$$\text{Total extraction yield} = \frac{\text{wt. of (dry coal-residue)}}{\text{wt. of d.a.f. coal}} \times 100 \; (\%)$$

Table III

Extraction of Coals in Anthracene Oil

Digestion conditions 400°C,
60 minute residence time
Coal:solvent ratio 1:3

Coal	NCB coal rank code (CRC)	Characterisation of products 2		
		% gas	% coal in solution	% total extraction
Cynheidre beans	101	<1	6	6
Penalta	202	<1	3	3
Garw	204	1	46	47
Tilmanstone	204	2	34	36
Tymawr	204	2	39	41
Beynon	301a	1	67	68
Windsor	301a	1	70	71
Marine	301a	2	74	76
Oakdale	301a	2	70	72
Cwm (a)	301a	1	43	44
(b)	301a	1	50	51
(c)	301a	1	50	51
(d)	301a	1	41	42
Bedwas	301b	2	52	54
Herrington	401	2	80	82
Bersham	502	3	82	85
Wearmouth	502	3	76	79
Hawthorn	502	3	79	82
Manton	502	3	78	81
Barrow	502	2	76	78
Brodsworth (a)	602	2	75	77
Swallowood	602	3	60	63
Brodsworth (b)	702	3	60	63
Annesley	702	3	68	71
Rufford (a)	702	3	55	58
Rufford (b)	702	2	76	78
Blidworth	702	3	71	74
Markham	702	4	53	57
Whitwell	802	4	51	55
Bestwood high	902	4	45	49
Linby (aged)	902	4	45	49
Linby (fresh)	802	4	54	58

Note 2 d.a.f. basis; on weight of coal

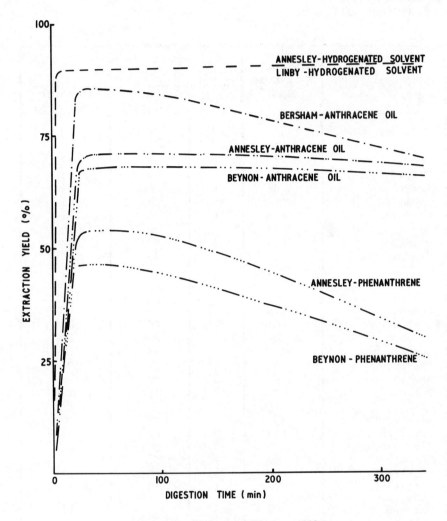

Figure 4.   Extraction of coals at 400°C

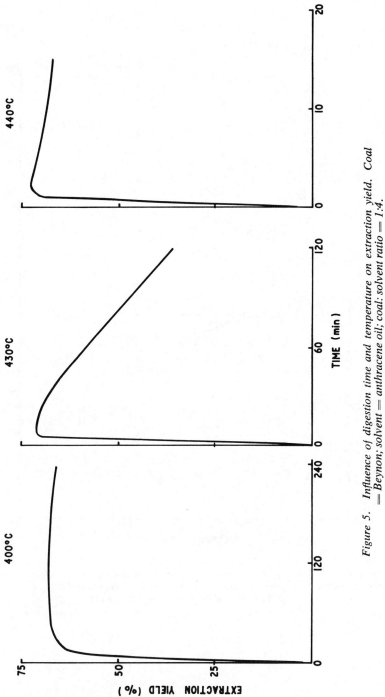

Figure 5. Influence of digestion time and temperature on extraction yield. Coal = Beynon; solvent = anthracene oil; coal: solvent ratio = 1:4.

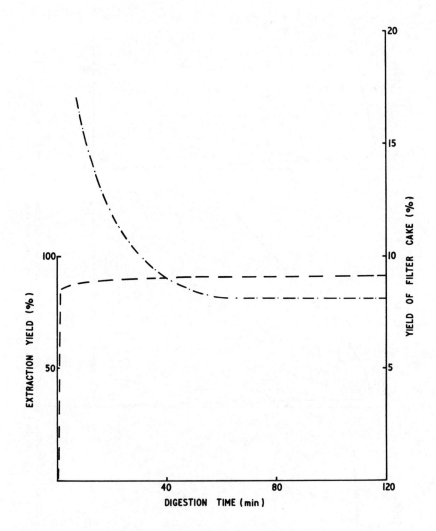

Figure 6.   Influence of digestion time on the extraction yield and yield of filter cake. Digestion: Annesley coal/hydrogenated solvent; prepared at 430°C: (— · —), yield of filter cake; (— — —), extraction yield.

Discussion of Results

   Influence of Digestion Conditions. To make a comparison
between the reactivity of coals it was necessary to select the
digestion conditions under which the maximum extraction yield
is obtained. For digests prepared in anthracene oil or
phenanthrene at 400°C the maximum extraction yield was obtained
with a residence time of 60 minutes (Figure 4). In comparison,
the extraction yields of coals in hydrogen donor solvents were
relatively insensitive to changes in residence time due to the
suppression of the polymerisation of the dissolved coal by
hydrogen transfer from the solvents. The extraction yield is
also dependent upon the digestion temperature and it has been
shown that a relationship exists between the digestion time and
temperature (14). A shorter digestion time at a relatively
high temperature can be equated to a longer digestion at a lower
temperature (Figure 5).
   Where digests are prepared with short residence times it is
important to consider the physical changes occurring in the coal
particles during digestion (8, 14). The mechanism operating is
not fully understood but it has been suggested that in the early
stages of digestion the coal particles contain some
depolymerising gel-like material which is indistinguishable from
the dissolved coal by quinoline solubility. A measure of the
amount of depolymerised coal retained in the coal particles has
been made from the weight of filter cake after filtration of
digests prepared using different residence times. To simplify
the interpretation of the results it was convenient to prevent
the polymerisation of the dissolved coal by use of a hydrogen
donor solvent. The digests were prepared at 430°C to facilitate
rapid filtration. The results show a decreasing yield of filter
cake with digestion time up to 60 minutes, whilst little
variation in the extraction yield as measured by quinoline
solubility is observed (Figure 6). In view of the length of
time required for dissolution of the coal a standard digestion
of 400°C and 60 minutes residence time was used for comparison
of coals.
   It has been shown that the coal to solvent ratio has
little influence upon the extraction yields when using coal
concentrations in the range 10 to 50% (15).

   Selection of Solvents. The extraction yield of a low rank
coal (Annesley) has been determined after digestion using a
selection of solvents (Table II). The results show large
variations in solvent power and, in particular, the high
extraction yields obtained with hydrogen donor solvents. It is
important to differentiate between the ability of a solvent to
prevent polymerisation of the dissolved coal by hydrogen
transfer, and its ability to retain the dissolved coal in
solution. For example, Tetralin is frequently quoted as an

excellent solvent for coals, but if the digest is allowed to
cool a large proportion of the depolymerised coal is
precipitated. Anthracene oil does not suffer from this defect,
but a small amount of precipitate may form in digests prepared
using an hydrogenated anthracene oil depending on the process
conditions.

Anthracene oil or an hydrogenated process oil are the
solvents which will be used on a commercial plant. Hydrogen
donor solvents give consistently high extraction yields of the
coals irrespective of type and thus are unsuitable for the
comparison of coal reactivity. The lower solvent power of
anthracene oil would appear more suited to this purpose.
Unfortunately, changes in composition of the solvent between
batches gives rise to small differences in the extraction yield
of a coal. In particular, the small concentration of labile
hydrogen donor compounds (e.g. 9, 10 dihydroanthracene) will
have a significant effect. For this reason a pure low power
solvent, phenanthrene (the main component of anthracene oil)
was also selected as a solvent suitable for comparison of coal
reactivity.

Characterisation of Coals. The results from the extraction
of a range of coals in anthracene oil and phenanthrene (Tables
III and IV) show that the mid rank coals (CRC 401 and CRC 502,
Figure 2) are the most reactive. In comparison, the high rank
coals (CRC 101 and CRC 202) are almost totally inert. A
convenient graphical method of displaying the changes in
extraction yield is shown in Figure 7 where the data has been
superimposed on Seyler's coal chart. A similar classification
can be made using phenanthrene as solvent but with lower
overall extraction yields. Small increases in the gas yield
are observed with a decreasing rank of coal.

Influence of Maceral Composition. Coals of a similar
elemental composition can give different extraction yields.
For example, Beynon and Cwm coals when digested in anthracene
oil give extraction yields of 68% and 47% respectively. This
variation can be explained by reference to the maceral
composition of the coals. Beynon coal contains a lower concentration of inertinite than the Cwm coal (Table V). In experiments where relatively pure samples of petrographic species were
digested in anthracene oil, exinite and vitrinite were shown
to be highly soluble, whilst in comparison the inertinite was
almost completely insoluble. Similar variations in reactivity
of macerals have been reported from studies of solubility in
pure organic solvents (16).

A more detailed study has shown that the solubility of
vitrinites is dependent upon coal rank (17). Those from the
prime coking coals and mid rank coals (e.g. CRC 301a and

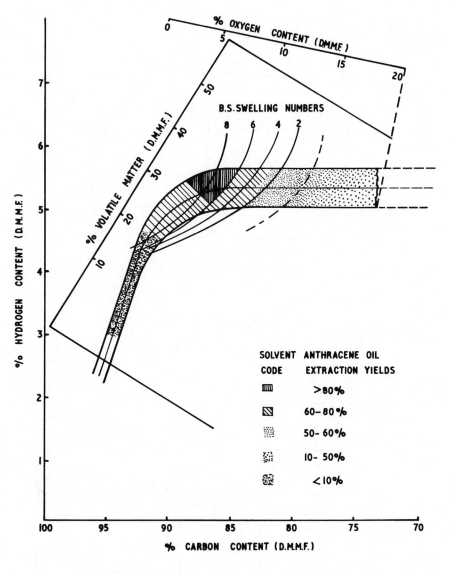

Figure 7.   Relationship between coal type and extraction yield

Table IV

Extraction of Coals in Phenanthrene

Digestion conditions 400°C
60 minute residence time
Coal:solvent ratio 1:3

Coal	NCB coal rank code (CRC)	Characterisation of products [2]		
		% gas	% coal in solution	% total extraction
Garw	204	1	14	15
Beynon	301a	2	44	46
Oakdale	301a	2	43	45
Bersham	502	2	78	80
Annesley	702	4	51	55
Linby	902	4	24	28

Table V

Influence of Maceral Composition

Digestion conditions 400°C,
60 minute residence time
Coals: selected from NCB rank CRC301a
Solvent: anthracene oil

Coal	Total extraction [1] yield (%)	Maceral Composition (%)		
		Vitrinite	Exinite	Inertinite
Beynon	68	78	0.6	17
Windsor	71	67	2.0	29
Marine	76	71	4.0	24
Oakdale	72	73	0.2	24
Cwm (a)	44	58	0.6	38
(b)	51	68	0.2	30
(c)	51	59	0.6	37
(d)	42	64	0.4	35

CRC 502) are the most soluble. In comparison, the extraction yields of exinites were independent of coal rank. The subdivision of inertinites into fusinite and micrinite showed that fusinite was completely insoluble in anthracene oil, whilst the micrinite was slightly soluble.

The petrological composition is important when considering the solvent extraction of prime coking coals but with lower rank British coals the variations in petrology are less pronounced. A more frequent cause of variations in extraction yield with low rank coals (CRC 802 and CRC 902) results from ageing. The reactivity of a coal decreases substantially as the coal becomes oxidised by exposure to the atmosphere (Table III).

Application to Larger Scale Plant. To determine whether this method of coal classification is applicable on a larger scale, a series of comparative tests were made using a 200 litre autoclave and a continuous digester with a throughput of 120 kg h^{-1}. The results generally show good agreement between the extraction yields from digests prepared in the maxibomb (as determined from quinoline solubility) and data from mass balance on larger scale digesters (Table VI).

Blending of Coals. In practice it is unlikely that a coal from a single source will be used as a feed to a large solvent extraction plant and a blend of several coals may have to be used. A preliminary series of tests using blends gave extraction yields in reasonable agreement with those derived from extraction of individual coals (Table VII).

Conclusion

A system based upon the reactivity of coals during extraction with anthracene oil and phenanthrene has been developed. A convenient graphical method of expressing the data on Seyler's chart has been adopted. This method has limitations when dealing with prime coking coals, which show wide variations in extraction yield. The differences in extraction yield relate to the concentration of inertinite which is virtually insoluble in anthracene oil.

It has been shown that data from this classification is applicable to large scale digesters and can be used with blends of different coals.

Acknowledgement

This paper is published with the permission of the National Coal Board. The views expressed in this paper are those of the authors and not necessarily those of the Board.

Table VI

Comparison of Extraction Yields Between Different Digesters

Digestion conditions nominally
400°C with 60 min residence time
Solvent: anthracene oil

Coal	Extraction yield[1] (%)		
	Maxibomb	200 litre autoclave	2 tonne per day extract plant
Garw	47	52	45
Beynon	68	68	63
Bersham	85	78	78
Oakdale	72	70	56
Manton	81	72	–
Whitwell	55	56	–

Table VII

Influence of Coal Blend on Extraction Yields

Digestion conditions 400°C
60 minute residence time
Solvent: anthracene oil
Coal:solvent ratio 1:3

Coals		Ratio A:B	Extraction yield[1] (%)	
A	B		As measured	calculated
Garw	Bersham	25:75	68	75
		50:50	65	66
		75:25	61	57
	Annesley	50:50	55	60
Oakdale	Annesley	50:50	66	72
	Manton	50:50	71	76
	Manton	75:25	68	74

Abstract

A system of classifying coals for solvent extraction, based upon the extent of extraction when using anthracene oil and phenanthrene as solvents has been developed. The reactivity of the coals can be conveniently presented by superimposing the results on Seyler's coal chart. The effects of variations in maceral composition are also discussed.

It has been shown that the results obtained are applicable to large scale reactors and to blends of coals.

Literature Cited

1. National Coal Board, Annual Report (1977/78).
2. Seyler, C.A.; Proc. S. Wales Inst. Engr., 1938, 53, 254 and 396.
3. Falkum, F.; Glenn, R.A.; Fuel, 1952, 31, 133.
4. Van Krevelen, D.W.; "Coal", Elsevier, Amsterdam, 1961.
5. Berkowitz, N.; Proc. Symposium of the Sci. and Tech. of Coal, Ottawa, Department of Energy, Mines and Resources (Canada), 1967, 149.
6. Oele, H.P.; Waterman, H.I.; Goedkoop, M.L.; Krevelen, D.W.; Fuel, 1951, 30, 169.
7. Curran, G.P.; Struck, R.T.; Gorin, E.; Preprint, Div. Petroleum Chem., Am. Chem. Soc.; 1966, 130.
8. Curran, G.P.; Struck, R.T.; Gorin, E.; Ind. Eng. Chem. Process Des. and Dev., 1967, 6, 166.
9. Neavel, R.C.; Fuel, 1976, 55, 237.
10. Liekenberg, B.J.; Porgieter, H.G.J.; Fuel, 1973, 52, 131.
11. Gun, S.R.; Sama, J.K.; Chowdhury, P.B.; Mukherjee, S.K.; Mukherjee, D.H.; Fuel, 1979, 58, 171 and 176.
12. Rantell, T.D.; Clarke, J.W.; Fuel, 1978, 57, 147.
13. Spiers, H.M., Technical Data on Fuel, London, British National Committee World Power Conference, 1961.
14. Clarke, J.W.; Rantell, T.D.; Paper submitted to Fuel.
15. Rantell, T.D.; Clarke, J.W.; Unpublished data.
16. Lowry, H.H.; Chemistry of Coal Utilisation, Supplementary volume, Wiley, New York, 1963.
17. Kimber, G.M.; Unpublished data.

RECEIVED March 28, 1980.

NOVEL LIQUEFACTION PROCESSES

A New Outlook on Coal Liquefaction Through Short-Contact-Time Thermal Reactions: Factors Leading to High Reactivity

D. DUAYNE WHITEHURST

Mobil Research and Development Corporation, Princeton, NJ 08540

About 35 years ago German investigators observed that the initial phases of coal liquefaction in presence of hydrogen donors involved the conversion of the insoluble coal matrix into a form which is soluble in strong organic solvents such as pyridine (1). Work in the United States by Gorin (2) and Hill (3) showed that such transformations are extremely rapid and require the consumption of relatively little hydrogen from the solvent. These initial products are highly functional molecules having molecular weights of 300-1000 and become soluble in weaker solvents such as benzene or hexane only after the degree of functionality and molecular weight are reduced (4). They have been referred to by a variety of names but in this paper they will subsequently be called asphaltols.

Conversion of coal to benzene or hexane soluble form has been shown to consist of a series of very fast reactions followed by slower reactions (2,3). The fast initial reactions have been proposed to involve only the thermal disruption of the coal structure to produce free radical fragments. Solvents which are present interact with these fragments to stabilize them through hydrogen donation. In fact, Wiser showed that there exists a strong similarity between coal pyrolysis and liquefaction (5). Recent studies by Petrakis have shown that suspensions of coals in various solvents when heated to ∿450°C produce large quantities of free radicals (∿.1 molar solutions!) even when subsequently measured at room temperature. The radical concentration was significantly lower in H-donor solvents (Tetralin) then in non-donor solvents (naphthalene) (6).

The production of such high concentrations of radicals leads to a very unstable situation and if the radicals are not stabilized via H-donation, they undergo a variety of undesired reactions such as condensation, elimination or rearrangement (7). Neavel has shown that at short times (∿5 min) a vitrinite enriched bituminous coal can be converted to ∿80% pyridine soluble form in even non-donor reaction solvents (naphthalene) (8). But if reaction times are extended, the soluble products revert to an insoluble form via condensation reactions. Such condensation reactions were

0–8412–0587–6/80/47–139–133$08.00/0

proposed to involve hydrogen abstraction from hydroaromatic coal structure (7,8). Indeed, coals have been shown to contain large quantities of labile hydrogen (9) and in solvents containing limited H-donors, coal products are more aromatic than those in H-donor rich solvents. Figure 1 shows the hydrogen consumption measured for a series of conversions of a bituminous coal (Illinois #6 - Monterey Mine) in which solvents of varying H-donor content were used. It can be seen that the hydrogen required to produce 450-°C products and lower heteroatom contents were essentially the same for all solvents but as the solvent H-donor content was decreased, H_2 gas and SRC provided the needed hydrogen.

The significance of the above-described work is that in all of the presently developing coal liquefaction processes, the initial step in the conversion is thermal fragmentation of the coal structure to produce very fragile molecules which are highly functional, low in solubility, and extremely reactive toward dehydrogenation and char formation. A more detailed discussion of the chemical nature of these initial products has been presented elsewhere (4).

The formation of these thermal fragments is necessary to catalytic liquefaction processes before the catalysts can become effective for hydrogen introduction, cracking and/or heteroatom removal (10).

In thermal processes the formation of asphaltols always precedes other reactions such as major heteroatom rejection and distillate formation. In fact, in the SRC process bituminous coals are actually dissolved by the time the coal slurry exits the preheater (4,11). This has recently been demonstrated at the SRC process development unit (PDU) in Wilsonville, Alabama (11) (see Figure 2).

These observations suggest that new coal liquefaction technology may be possible based on short contact time reactions. The purpose of this and the related papers in this volume by R.H. Heck and W.C. Rovesti is to show some potential advantages for optimized or integrated short contact time liquefaction processes over conventional technology.

This paper will concentrate on factors which lead to high conversion at short time. R.H. Heck, T.O. Mitchell, T.R. Stein and M.J. Dabkowski discuss the relative ease of conversion of short and long contact time SRCs to higher quality products. C.J. Kulik, W.C. Rovesti and H.E. Liebowitz discuss some new leads presently being explored at the Wilsonville PDU in which short contact time liquefaction is being coupled with rapid product isolation via the Kerr-McGee Critial Solvent Deashing Process.

Advantages for Short Contact Time Coal Liquefaction

In order to understand the potential advantages for short contact time liquefaction processes, let us first consider some of the disadvantages for presently developing long contact time processes. These are enumerated below.

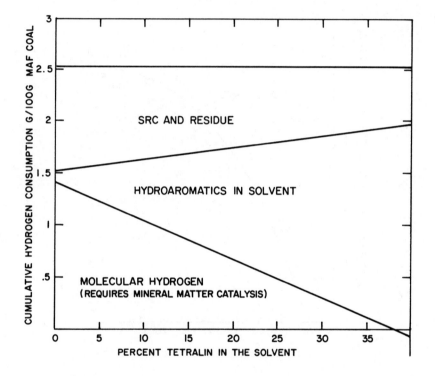

Figure 1.   The source of hydrogen is controlled by the solvent

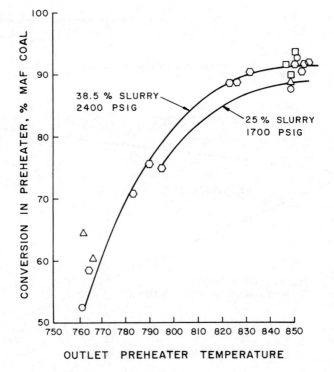

Figure 2.   *Effect of temperature on coal conversion: (○), 600–650 lb/hr; (□),*
300–350 lb/hr; (△), 730–830 lb/hr; (◇), 400–450 lb/hr.   Gas rate = 0–10,000
scfh

- Long contact time thermal processes have the intrinsic disadvantage of poor selectivity for light hydrocarbon gas formation relative to heteroatom removal (see Figure 3).

- Some desulfurization occurs thermally but essentially no denitrogenation occurs without the aid of catalysts.

- In long contact time thermal processes, essentially no net hydrogen is introduced into the heavy liquid products and the major product (SRC) continually dehydrogenates with increasing time (4,11). These last two points are illustrated in Table I and Figure 4.

- In catalytic coal liquefaction processes, reaction temperatures must be high in order to insure that thermal reactions disrupt the coal structure to the point that the catalyst can act on the products. As a result, selectivity is not optimal and excessive hydrocracking results (10). Catalyst aging is also excessive.

- Poor hydrogen utilization results in less than optimal thermal efficiencies for all developing processes.

TABLE I

HETEROATOM REMOVAL

	General Formula	Number Heteroatoms/100 C
Monterey Coal (maf)	$C_{100}H_{88}N_{1.6}O_{13.2}S_{1.74}$	14.9
Monterey SRC, short contact time (AC-59)	$C_{100}H_{88}N_{1.6}O_{9.1}S_{1.3}$	12.0
Monterey SRC, long contact time (AC-58)	$C_{100}H_{84}N_{1.7}O_{5.1}S_{0.7}$	7.5

To improve selectivity and conservation of hydrogen over present liquefaction technology in the conversion of coal to high quality liquids, we believe that thermal reactions should be kept as short as possible. Catalytic processes must be used for upgrading but should be used in a temperature regime which is optimal for such catalysts.

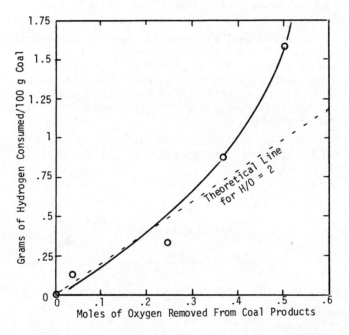

Figure 3.   Sensitivity of hydrogen consumption to oxygen removal

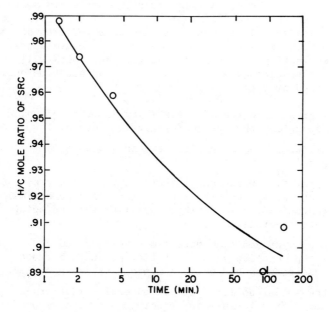

Figure 4.   H/C mole ratio of coal products vs. time

Conversion levels should be limited so as to be compatible with the required hydrogen manufacture from unconverted coal.

To illustrate this latter point, Figure 5 shows the calculated amount of coal required for hydrogen manufacture as a function of the rank of the starting coal and the composition of the desired products (10). In these calculations a 12.5% methane byproduct was assumed and the thermal efficiency of the hydrogen generation was assumed to be 70%.

The significance of these calculations is that lower rank coals will require ∿5% lower conversion than higher rank coals for a given end product. Also, the more severe a coal is to be upgraded, the lower its conversion has to be in the initial phases of liquefaction. One very pertinent question to be addressed is whether or not coals can be converted to the levels shown in Figure 5 in a short contact time process. This paper will deal with that question as well as what compositional features of the coal and the solvent influence short contact time conversions.

A combination of this and the related papers will show the following potential advantages for short contact time optimized or integrated processes.

- A greater degree of flexibility is achieved by decoupling thermal and catalytic processes.

- Higher yields of desired liquid products are possible.

- Catalytic upgrading of short contact time products allows catalysts to be used in more optimal conditions; selectivities are improved and aging rates decreased.

- Less hydrogen consumption may be required for the production of high quality products because less gaseous hydrocarbon byproducts are produced.

- Low sulfur boiler fuel can potentially be produced from low sulfur western coals with reduced capital investment.

The Effect of Coal Composition on Short Contact Time Conversion

The classic work of Storch and co-workers showed that essentially all coals below ∿89% C_{maf} can be converted in high yields to acetone soluble materials on extended reaction (12). We have investigated the behavior of coals of varying rank toward short contact time liquefaction. In one series of experiments, coals were admixed with about 5 volumes of a solvent of limited H-donor content (8.5% Tetralin) and heated to 425°C for either 3 or 90 minutes. The solvent also contained 18% p-cresol, 2% γ-picolene, and 71.5% 2-methylnaphthalene and represented a synthetic SRC recycle solvent. The conversions of a variety of coals with this

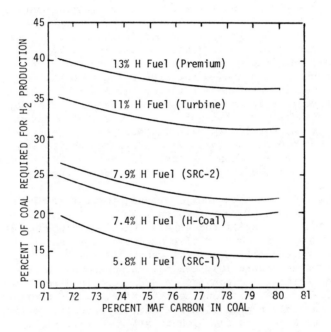

Figure 5.   Hydrogen requirement for conversion of coal

solvent to pyridine soluble materials are shown in Figure 6. As
in the work of Storch (12), we observed that coal conversion at
long times was high for all coals having less than 88% C_{maf}. At
short contact time, however, both low and high rank coals were not
converted in high yield to pyridine soluble form. Only coals from
77 to 87% C_{maf} were converted to more than 70% in 3 minutes. The
lack of conversion for low rank coals led to an investigation of
what compositional features of the coal had limited their conversion.

This investigation showed that although the low rank coals
did not produce as much pyridine soluble products, they had indeed
undergone major compositional change. This was evidenced by high
hydrogen consumption and the production of large quantities of CO_2
as shown in Figure 7. The reflectivity of the unconverted solids
was much higher than that of the original coals as shown in Table
II. This also indicates a major compositional change.

An oxygen balance calculation of the products of short contact time conversion showed that the percent loss of oxygen from
the coal was also much more advanced for low rank coals than for
high rank coals. Figure 8 illustrates the percent oxygen conversion for a variety of coals at short times (2-5 minutes). If this
fraction of the total oxygen is compared to the fraction of the
total oxygen present as carboxyl (13) and carbonyl groups (14), an
almost 1 to 1 correlation results. The average distribution of the
various oxygen-containing functional groups in coal is shown in
Figure 9 (15).

It is believed that a major reason for the insolubility of
short contact time products of low rank coals is that the insoluble materials are still too highly functional (phenolic) to be
soluble. The short contact time SRCs from low rank coals do indeed contain less polyfunctional materials than the SRCs of high
rank coals (see Figure 10). We have also shown that for low rank
coals, the initial low conversion SRCs contain a lower proportion
of phenolic oxygen than long contact time, high conversion SRCs
(16). A possible explanation is that the phenolic content of low
rank coal products are too high to allow solubility in even pyridine. This point has yet to be proven, however.

In addition to functionality, skeletal structure and the
physical make-up of the coal was found to be important in achiev-
ing high conversions at short time. Neavel has previously called
attention to the importance of plasticity in coal liquefaction
(17). Mochida and co-workers have also shown that the degree of
solubilization of coals in polyaromatic solvents relates directly
to their fluidity (18). We did not obtain fluidity measurements
directly on this series of coals but data developed by Honda (19)
indicate that a maximum in fluidity occurs at ∿85% C_{maf} (see
Figure 11). This is in the same region of carbon content in
which we observed a maximum in conversion at short time (see
Figure 6).

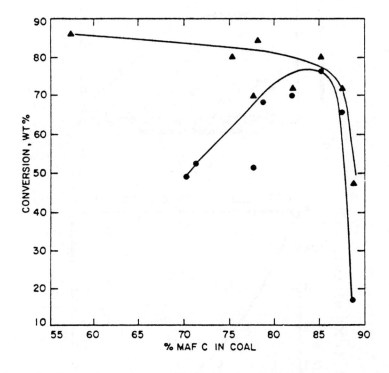

Figure 6.   Conversion vs. percent MAF after 3 and 90 min: (●), 3-min ring; (▲), 90-min ring.

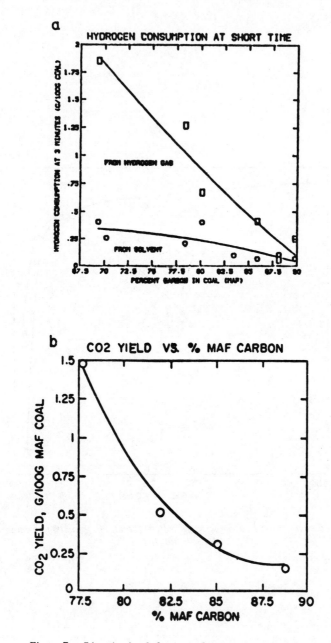

Figure 7.   Liquefaction behavior at 3 min as a function of rank

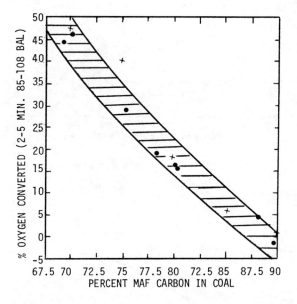

Figure 8.   Percent oxygen converted vs. rank (2–5 min): (●), percent oxygen lost as CO or CO₂; (✕), percent oxygen in coal as carboxyl or carboxyl groups.

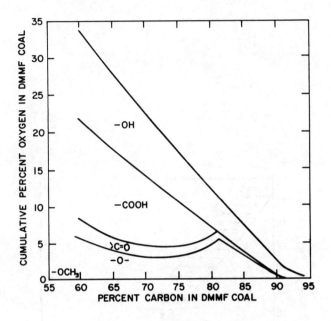

Figure 9.   Distribution of oxygen functionality in coals

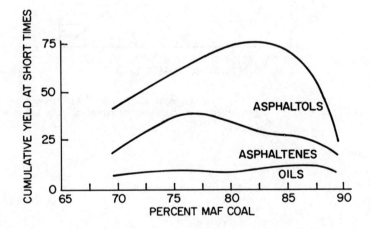

Figure 10.   SRC composition at short time

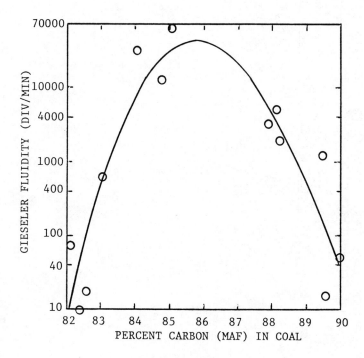

Figure 11.   Maximum fluidity vs. rank of coal

TABLE II

COMPARISON AMONG RANK, CONVERSION AND REFLECTANCE
OF HIGH VITRINITE COALS AND RESIDUES

Mobil Run No.	PSOC	Rank	% Conversion	Reflectance* of Vitrinite in Coal	Reflectance* of Vitroplast in Residue	% Change in Reflectance
AC-152	312	hvCb	51	0.44	0.97	+120
AC-151	330	hvBb	70	0.76	1.36	+79
AC-150	372	hvAb	79	1.00	1.23	+23
AC-149	256	mvb	65	1.26	1.46	+16
AC-148	405	lvb	17	1.68	1.64	- 2

*Percentage of incident light, in oil.

Another parameter is the intrinsic extractability of the parent coals by pyridine. As can be seen in Figure 12, the shape of the curve of pyridine extract yield from the various coals vs. their carbon content follows the same trend as the short contact time conversions of these coals.

It has been proposed (17) that the portion of coal which is mobile under liquefaction conditions, contributes to the stabilization of thermally-generated radicals. Thus, coals which are highly fluid or contain large contents of extractable material might be expected to provide hydrogen and thus promote conversion. Collins has reported that vitrinite is a better donor of hydrogen than is Tetralin (20). Our own measurements of the aromatic content and elemental analyses of the coals (16,21) (or coal products) before and after conversion at short time are insufficient to confirm or deny the supposition that coal acts as its own H-donor even at short times.

There is a clear trend, however, in the content of aromatic carbon in a coal and its convertibility at short times. This is shown in Figure 13. It can be seen that high convertibility occurs for coals which are intermediate in aromatic carbon content. This observation is consistent with the common belief that thermal fragmentation occurs at aliphatic positions α or β to aromatic rings. If the aromatic content becomes too high, the concentration of such aliphatic linkages must become limited.

Working co-operatively with others, we have found some indication that certain alilphatic linkages between aromatic nucleii are involved in the rapid dissolution of coal. The absolute aliphatic hydrogen content as determined by P. Solomon using FTIR (22) shows a very good linear relationship with conversion of coal in 3 minutes to pyridine soluble materials (Figure 14a).

Deno has also developed an analytical procedure for determining the type and amount of aliphatic constituents in coals and natural products (23). This procedure selectively oxidizes aromatic nucleii and does not attack saturated aliphatic structures. Among the structures which can be identified are Ar-CH$_2$-CH$_2$-Ar and Ar-CH$_2$-aliphatic. Such structures could constitute some easily broken C-C bonds in coals. A limited number of coals were oxidized and the amount of hydrogen of the types identified above were determined. These results were compared with the yield of SRC achieved in short contact time conversions. Figure 14b shows that there is a rough correlation between SRC yield and certain aliphatic structures. These encouraging initial results do suggest that further work in this area could help in understanding the nature of the reactive aliphtic structures in coal.

These results indicate that the aliphatic portion of the coal is very important in the initial phases of coal conversion. Weak linkages must be associated with the aliphatics in coal though they have not as yet been completely identified. Both of the above methods show an increase in the aromatic methyl content of SRCs at short times which indicates that cleavage at a benzylic carbon is important in dissolving the coal.

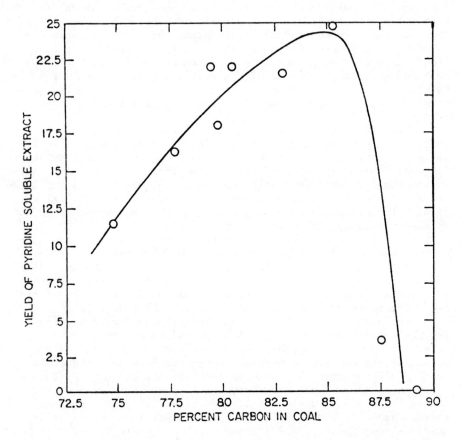

Figure 12. Yield of extract (percent weight recovered) vs. rank (percent MAF carbon)

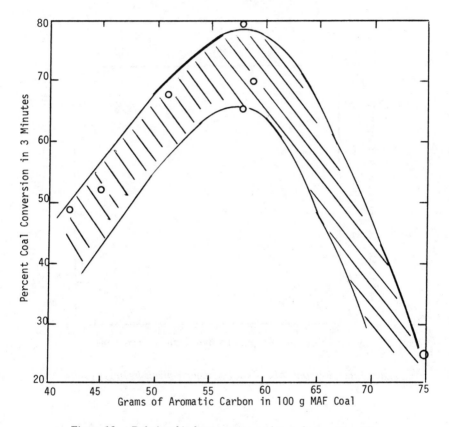

Figure 13.   Relationship between conversion and aromatic carbon

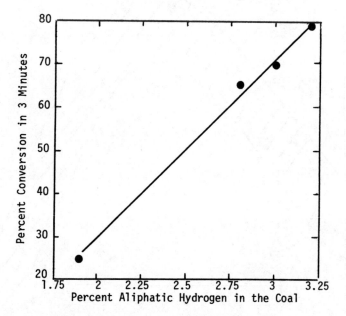

Figure 14a.   Response of coal conversion to aliphatic hydrogen content

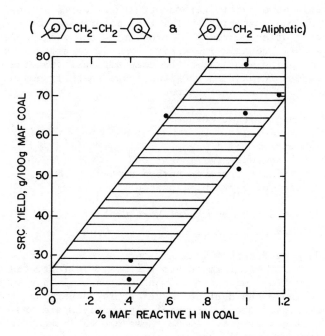

Figure 14b.   *Relationship between SRC yield and reactive aliphatic hydrogen*

There is additional evidence for the importance of aliphatic hydrogen and its relationship to coal rank and reactivity in lique- faction. Reggel, Wender, and Raymond (9) studied the dehydro- genation of vitrains from a variety of coals with 1% Pd/CA(CO₃)₂ in refluxing phenanthridine. Coals in the rank range which rapidly give high SRC yields are rich in hydrogen, which their technique can remove. Furthermore, there was a distinct difference between bituminous coals, subbituminous coals and lignites. The lower rank materials yielded less H_2 in their test; we find these to be very reactive but slow to yield pyridine-soluble products. These work- ers concluded from their work "that lignites and subbituminous coals contain some cyclic carbon structures which are neither aro- matic nor hydroaromatic; that low rank bituminous coals contain large amounts of hydroaromatic structures; and that higher rank bituminous coals contain increasing amounts of aromatic structures".

The following summarizes the compositional features of coals which have been identified as significant to their potential con- vertibility to SRC at short times:

- High vitrinite content
- High fluidity
- High extractability
- Carbon contents (maf) near 85%
- Intermediate aromatic carbon content
- Presence of certain aliphatic structures

Effect of Process Parameters on Short Contact Time Conversions

The data on Figure 6 indicate that some coals are difficult to convert to soluble form at short times. In fact, the degree of conversion at 425°C with the solvent chosen would not be high enough to balance the hydrogen manufacture/conversion stoichio- metry, shown in Figure 5. Several alternatives are available to increase this conversion. Among these are to increase the temper- ature and/or pressure of the reaction.

At present, our data are not definitive on the effect of in- creasing H_2 pressure. However, increasing the temperature has a profound effect. This was clearly demonstrated by early workers in the field (2,3). More recently it has been shown by Morita and Hirosawa that for a given coal there is a temperature above which conversion to soluble form no longer increases even at short time (24). Their data for one coal is shown in Figure 15. Kleinpeter and Burke have reported a similar result for a bituminous coal of the United States (25). Sensitivity to temperature is most prob- ably dependent on the nature of the solvent used for the conver- sion. This preliminary conclusion is based on the work of Neavel (8) and our own work with solvents having high H-donor contents (40% Tetralin). Figure 16 shows that raising the temperature to ∿450°C has no detrimental effect for 3 U.S. bituminous coals with this solvent.

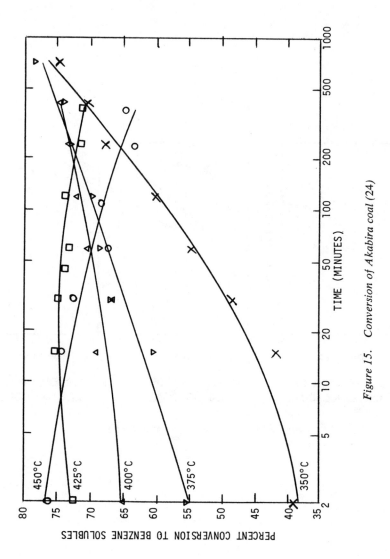

Figure 15.   Conversion of Akabira coal (24)

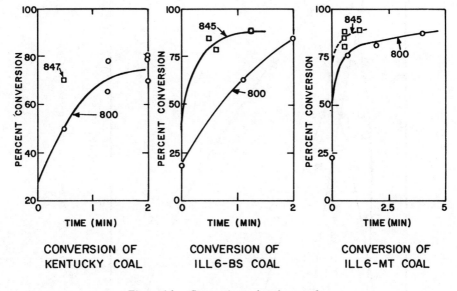

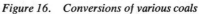

Figure 16.   Conversions of various coals

Similarly, a subbituminous coal (Wyodak-Anderson) which gave only 60% conversion at 427°C in 2 minutes could be converted to >70% at 460°C with no ill effects (see Figure 17). At ∿470°C there was an indication that conversion had begun to decline at ∿2 minutes; however, this data is extremely limited. The implications of these results with a western subbituminous coal is that a low sulfur boiler fuel may potentially be produced in a single-stage short contact time process.

Effect of Solvent Composition of Short Contact Time Conversions

As discussed above, the composition of the solvent used in short contact time conversions can be important. The concentration of H-donors is one factor to be considered. It is known that in long contact time conversions, solvents having high H-donor contents have a better ability to prevent char formation as sulfur is removed from the SRC. Thus, higher yields of upgraded liquids are observed when solvents containing high concentrations of H-donors are used.

If the initial reactions of coal are purely thermal, one might expect that the H-donor level will be of minor importance if times are kept short. In fact, all coals contain a certain portion of material that is extractable by pyridine. On heating coals to liquefaction temperatures, some additional material also becomes soluble in even non-donor solvents. Thus, there is a portion of all coals which can be solubilized with little dependence on the nature of the solvent.

Table III shows that hydrogenated and unhydrogenated SRC recycle solvents were equally effective for the conversion of a western subbituminous coal at low reaction severity. At higher severity but at times shorter than 10 minutes, significantly higher conversions were achieved only with the hydrogenated solvents which could donate more hydrogen.

TABLE III

SOLVENT EFFECTS ON SHORT TIME CONVERSION
OF BELLE AYR SUBBITUMINOUS COAL
(800°F, ∿3 min., 1500 psi H_2)

	Solvent	%H	Conversion
400–800°F	Recycle Solvent (Wilsonville)	8.15	59
400–800°F	Hydrogenated Recycle Solvent	9.67	58

We have observed that at short contact times the conversion of bituminous coals is also responsive to the level of H-donor in the solvent. Table IV shows the conversions of an Illinois #6

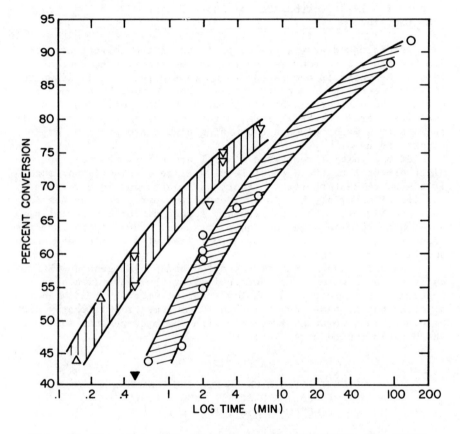

Figure 17.   Conversion of Wyodak coal with time: (○), 800°C; (□), 820°C;
(△), 840°–850°C; (▽), >860°C; (▼), 878°C.

bituminous coal which was heated for 2-3 minutes at 425°C in solvents of varying H-donor contents. The conversions increased from 50 to 85% conversion as the tetralin level was raised from 0% to 43% of the solvent.

TABLE IV

EFFECT OF SOLVENT* COMPOSITION OF CONVERSION OF
ILLINOIS #6 (BURNING STAR) COAL AT SHORT TIME
(2-3 minutes, 425°C)

% Tetralin in Solvent	% Conversion
0	50
8.5	68
43	85

*All solvents contained 2-methylnaphthalene as the major component. In some cases ∿18% p-cresol was also present.

Hydrogen donors are, however, not the only important components of solvents in short contact time reactions. We have shown (4,7,16) that condensed aromatic hydrocarbons also promote coal conversion. Figure 18 shows the results of a series of conversions of West Kentucky 9,14 coal in a variety of process-derived solvents, all of which contained only small amounts of hydroaromatic hydrocarbons. The concentration of di- and polyaromatic ring structures were obtained by a liquid chromatographic technique (4c). It is interesting to note that a number of these process-derived solvents were as effective or were more effective than a synthetic solvent which contained 40% tetralin. The balance between the concentration of H-donors and condensed aromatic hydrocarbons may be an important criterion in adjusting solvent effectiveness at short times.

Kleinpeter and Burke have recently reported (24) that solvents can also be over hydrogenated and thus become less effective in short time processes. Figure 19 shows some of their work in which a process-derived SRC recycle solvent was hydrogenated to various severities and used for the conversion of an Indian V bituminous coal. The results clearly show a maximum at intermediate hydrogenation severities. Our assessment of this observation is that the loss in conversion was due primarily to the loss in condensed aromatic nucleii rather than conversion of hydrogen donors to saturates.

Summary

To summarize, we have identified a number of features unique to short contact time coal liquefaction. The important factors

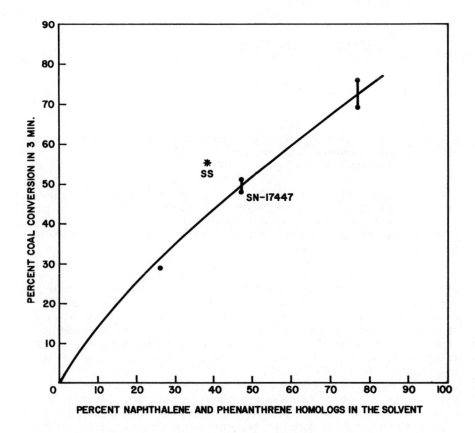

Figure 18.   Conversion of West Kentucky coal in various solvents

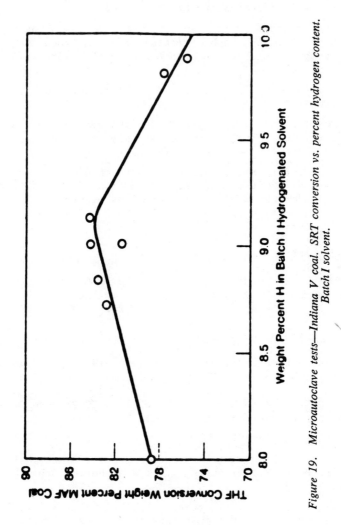

Figure 19.   Microautoclave tests—Indiana V coal. SRT conversion vs. percent hydrogen content. Batch I solvent.

which affect the application of short contact time conversions are tabulated below:

- Very high conversion is not necessary because of hydrogen manufacture requirements.

- Practically all coals can be converted to the desired stoichiometry at short time.

- High reactivity is associated with coals having high fluidity, high extractability, intermediate aromaticity, and the presence of weak aliphatic linkages.

- For each coal an optimal temperature for conversion exists.

- For a given coal a certain portion can be converted to soluble form very easily and is independent of solvent composition.

- Beyond this easily converted portion of the coal even at short times, the composition of the solvent is important - high concentrations of H-donors and polyaromatics are beneficial. Over hydrogenation is detrimental.

Acknowledgement

I would like to acknowledge EPRI and MRDC who funded the work and my co-workers F. J. Derbyshire, J. J. Dickert, M. Farcasiu, B. Heady, T. O. Mitchell, and G. A. Odoerfer.

Literature Cited

1. Rank, Report of Ludwigshafen, Germany (1942), U.S. Bureau of Mines T-346.

2. Curran, G.P.; Struck, R.T.; Gorin, E. I&EC Process Design and Development, 1967, 6, 166.

3. Hill, G.R. Fuel, 1966, 45, 326.

4. (a) Farcasiu, M.; Mitchell, T.O.; Whitehurst, D.D. Chem. Tech., 1977, 7, 680.

    (b) Whitehurst, D.D.; Farcasiu, M.; Mitchell, T.O., "The Nature and Origin of Asphaltenes in Processed Coals," EPRI Report AF-252, First Annual Report Under Project RP-410, February 1976.

    (c) Whitehurst, D.D.; Farcasiu, M.; Mitchell, T.O.; Dickert, J.J., "The Nature and Origin of Asphaltenes in Processed Coals," EPRI Report AF-480, Second Annual Report Under Project RP-410-1, July 1977.

5. Wiser, W., Fuel, 1968, 47, 475.

6. Petrakis, L.; Grandy, D.W., Fuel Div. Preprints, 146th ACS Meeting, Miami, Florida, November 1978, p. 147.

7. Whitehurst, D.D., "Relationships Between Recycle Solvent Composition and Coal Liquefaction Behavior", Proceedings of the EPRI Coal Liquefaction Contractors Meeting, Palo Alto, Calif., May 1978.

8. Neavel, R. C., Fuel, 1976, 55, 237.

9. Reggel, L.; Wender, I.; Raymond, R.,
   (a) Science, 1962, 41, 67.
   (b) Fuel, 1964, 43, 229.
   (c) Fuel, 1968, 47, 373.
   (d) Fuel, 1970, 49, 281.
   (e) Fuel, 1970, 49, 287.
   (f) Fuel, 1971, 50, 152.
   (g) Fuel, 1973, 52, 163.

10. Whitehurst, D.D.; Mitchell, T.O.; Farcasiu, M.; Dickert, J.J., Jr., "Exploratory Studies in Catalytic Coal Liquefaction", Final Report from Mobil Research and Development Corporation to EPRI under Project RP-779-18, 1979.

11. Lewis, H.E.; Weber, W.H.; Usnick, G.B.; Hollenack, W.R.; Hooks, H.W.; Boykin, R.G., Solvent Refined Coal Process Quarterly Technical Progress Report for the period January-March 1978, from Catalytic Inc. to EPRI and DOE, July 1978.

12. (a) Fisher, C.H.; Sprunk, G.C.; Eisner, A.; O'Donnell, H.J.; Clarke, L.; Storch, N.H., U.S. Bureau of Mines Technical Paper 642 (1942).

    (b) Storch, H.H.; Fisher, C.H.; Hawk, C.O.; Eisner, A., U.S. Bureau of Mines Technical Paper 654 (1943).

13. Schafer, H.N.S., Fuel, 1970, 49, 197.

14. Blom, L.; Edelhausen, L.; VanKrevlen, D.W., Fuel, 1957, 36, 135.

15. Whitehurst, D.D. The development of the correlations shown shown in Figure 9 will be published in the future.

16. Whitehurst, D.D.; Farcasiu, M.; Mitchell, T.O.; Dickert, J.J., Jr., "The Nature and Origin of Asphaltenes in Processed Coals," EPRI Report AF-1298, Final Report Under Project RP-410-1, December 1979.

17. Neavel, R.C., "Coal Plasticity Mechanism: Inferences from
     Liquefaction Studies", Proceedings of the Coal Agglomeration
     and Conversion Symposium, Morgantown, W.Va., May 1975 (pub-
     lished April 1976).

18. Mochida, I.; Takarabe, A.; Takeshita, K., Fuel, 1979, 58, 17.

19. Sanada, Y.; Honda, H., Fuel, 1966, 45, 295.

20. Collins, C. J.; Raaen, V. F.; Benjamin, B. M.; Kabalka, G. W.,
     Fuel, 1977, 56, 107.

21. Pines, A., "Aromatic Carbon Contents of Coals, SRCs, and
     Residues Were Determined by CP-^{13}C-NMR", Under a Subcontract
     to EPRI, RP-410-1.

22. (a) Solomon, P.R., "Relation Between Coal Structure and
     Thermal Decomposition Products", Preprints Fuel Division,
     ACS/CJS Chemical Congress, Honolulu, Hawaii, March 1979,
     p. 184.

     (b) Solomon, P.R.; Whitehurst, D.D., to be published.

23. Deno, N.C.; Greigger, B. A.; Jones, A.D.; Rakitsky, W.G.;
     Stroud, S.G., "Coal Structure and Coal Liquefaction",
     Final Report to EPRI, Project RP-779-16 (1979).

24. Morita, M.; Hirosawa, K., Nenryo Kyokai-Shi, 1974, 53, 263.

25. Kleinpeter, J.A.; Burke, F.P., Proceedings of the EPRI
     Contractors Conference on Coal Liquefaction, Palo Alto, CA,
     May 1979.

RECEIVED May 21, 1980.

Short-Residence-Time Coal Liquefaction

J. R. LONGANBACH, J. W. DROEGE, and S. P. CHAUHAN

Battelle, Columbus Laboratories, 505 King Avenue, Columbus, OH 43201

A two-step coal liquefaction process, which seems to have
potential to reduce hydrogen usage compared to conventional
solvent-refined coal (SRC) technology, has been investigated.[1]
The first step of the two-step process consists of a relatively
low temperature, low pressure reaction of coal with a coal-
derived solvent at short contact time in the absence of molecular
hydrogen. After removal of unreacted coal and mineral matter,
the second step is a short contact time, high pressure and temperature reaction of the soluble products in the presence of
molecular hydrogen. The purpose of the first step is primarily
to dissolve the coal and the purpose of the second step is to
reduce the sulfur level of the product and regenerate the solvent.
Decreased hydrogen consumption should result from the short contact times and the removal of mineral matter and pyritic and sulfate sulfur before the addition of molecular hydrogen.

Four variables were studied in the part of the experimental
program which examined the first step of the proposed two-step
process. The variables were: reaction temperature (413-454 C),
solvent to coal ratio (2:1 and 3:1), residence time (0-5 minutes),
and pressure (300-1800 psi nitrogen). Four experiments were done
to simulate the second step, in which hydrogenated solvent and
molecular hydrogen would be used to lower the sulfur content of
the product. These experiments were done at 441 C for 2 minutes,
with and without molecular hydrogen and recycle solvent containing
25 weight percent Tetralin.

Experimental

The experimental apparatus consisted of a 1-liter, stirred
autoclave (AC1), used to preheat the solvent-coal slurry,
connected to a 2-liter, stirred autoclave (AC2), equipped with
an internal heating coil to bring the solvent-coal slurry rapidly
to a constant reaction temperature, and a third autoclave (AC3),
equipped with a cooling coil to act as a quench vessel (Figure
1). This allowed direct determination of the material lost in

0–8412–0587–6/80/47–139–165$05.00/0
© 1980 American Chemical Society

each transfer step and an unambiguous determination of the
product yield. When the experiments required a high pressure
hydrogen atmosphere, a thermostatted hydrogen cylinder equipped
with a precise pressure gauge was used to determine the amount
of hydrogen gas before reaction. After reaction the autoclaves
were vented and the volume of gas was determined using a wet-test
meter and the percentage of hydrogen was determined by GC analysis of the gas samples. The slurry transfer lines were heat-
traced, 1/4-inch tubes equipped with quick-opening valves which
could be manually operated through a safety barrier. AC2 was
equipped with a thermocouple in the autoclave body as well as one
in the solution for precise temperature control during heating.
During venting the gases passed through a trap to condense
liquids, a gas sample port, an H_2S scrubber and a wet-test meter
to measure H_2S-free gas volume.

The experimental procedure consisted of preheating the coal-
solvent slurry to 250 C in AC1 while AC2 was heated empty to
slightly higher than the desired temperature. The slurry was
transferred to AC2 and the internal heater was used to bring the
slurry rapidly to constant reaction temperature. Typically,
heatup required 3.4 ± 0.6 min and the temperature remained constant within ±0.1 degree during the reaction period. The heatup
time was long in an apparatus of this size, but only 1-2 minutes
of the heatup period were spent at temperatures where the liquefaction rate is significant. After the reaction was complete,
the slurry was transferred to AC3 where it was quenched to 250 C
using an internal cooling coil. After the gases were vented, the
slurry was transferred to a heated filter and filtered at 250 C.

The product workup consisted of continuously extracting the
filter cake with tetrahydrofuran (THF) and combining the THF and
filtrate to make up a sample for distillation. In some experiments the THF extracted filter cake was extracted with pyridine
and the pyridine extract was included in the liquid products.
Extraction with pyridine increased coal conversion to soluble
products by an average of 1.6 weight percent. The hot filtrate-
THF-pyridine extract was distilled. Distillation cuts were made
to give the following fractions, THF (b.p. <100 C), light oil
(b.p. 100-232 C), solvent (b.p. 232-482), and SRC (distillation
residue, b.p. >482 C).

<u>Materials</u>

The two coals used for the experiments were blends of West
Kentucky 9 and 14 seam coals. One blend contained 0.39 weight
percent more organic sulfur than the other. Both coals were
significantly oxidized before they were received as shown by the
appreciable sulfate sulfur contents. Coal analyses are shown in
Table 1.

The solvent used for experiments simulating the first
process step was recycle solvent obtained from the Wilsonville

TABLE I. ANALYSES OF WEST KENTUCKY COAL 9/14 BLENDS (3)

(As-Received Basis)

Sample	I	II
Proximate Analysis, wt %		
Moisture	5.72[a]	2.71[b]
Ash	12.71	8.90
Volatile matter	34.7	36.9
Fixed carbon	47.5	51.49
Ultimate Analysis, wt %		
Carbon	66.0	70.0
Hydrogen	4.7	4.8
Nitrogen	1.3	2.4
Sulfur	3.75	3.16
Oxygen	12.2	10.74
Sulfur Types, wt %		
Total	3.77	3.16
Pyritic	1.96	1.30
Sulfate	0.60	0.31
Sulfide	0.01	--
Organic (by difference)	1.16	1.55

(a) Wilsonville SRC pilot plant sample No. 15793.

(b) Sample from Pittsburgh and Midway Coal Company.

SRC pilot plant. An analysis and distillation data for the
solvent are shown in Table 2. The solvent contained 5 percent
of material boiling below 232 C, the cutoff point between light
oil and solvent in the product distillation, and 4-5 percent of
material boiling above 482 C, the cutoff point between solvent
and SRC.

A blend of Wilsonville recycle solvent (75 weight percent)
and 1,2,3,4-tetrahydronaphthalene (25 weight percent) was prepared
for use as the solvent in experiments simulating the second
process step, which would use hydrogenated solvent. Analyses
and distillation data for this solvent are also given in Table 2.
Tetralin boils below 232 C and was collected in the light oil
distillation fraction during product workup.

Results and Discussion

THF Conversion. Tetrahydrofuran (THF) conversion was calculated from the difference between the initial and the final
solubilities of the total coal-solvent slurry in THF. It was
assumed that all of the solvent and none of the starting coal
was soluble in THF. THF conversions were calculated on an MAF
coal basis and adjusted for the coal not recovered from the autoclaves. The filter cake resulting from filtration of the product
at 250 C was continuously extracted with THF for up to 3 days.
The THF soluble conversion figures may be high however, since hot
recycle solvent is probably a better solvent for coal liquids
than THF and may have dissolved some material during the hot
filtration which would be insoluble in THF.

Figure 2 shows THF conversion plotted as a function of
reaction time and temperature at 3:1 solvent/coal ratio. The THF
solubles appear to be formed as unstable intermediates in the
absence of a good hydrogen donor. The low molecular weight THF
solubles may be able to combine in the absence of hydrogen to
form higher molecular weight materials which are insoluble in
THF. At 413 C, THF solubles increase between 0 and 2 minutes.
At 427 C, most of the THF solubles are produced during the heatup
period. THF solubles formation is over by the end of the heatup
period and is declining slowly after "zero" reaction time at
441 C. At 454 C the production of THF solubles is over before
the heatup period is completed and the decline is more rapid
than at 441 C.

Increasing the solvent to coal ratio might be expected to
have the effect of stabilizing the THF soluble materials by
making available more hydrogen from the solvent. However, the
changes in THF conversion as a function of solvent to coal ratio
at 1 minute residence time and 427-441 C are relatively small.
THF soluble conversion is increased by the presence of both
molecular hydrogen in the gas phase and Tetralin added to the
solvent.

TABLE II. SOLVENT ANALYSES (3) (a)

	Wilsonville Recycle Solvent	3:1 Wilsonville Recycle Solvent-Tetralin(b)
Ultimate Analysis, wt % as received		
Moisture	<0.01	Trace
Ash		0.10
Carbon	87.8	87.9
Hydrogen	7.8	8.3
Nitrogen	0.7	0.6
Sulfur	0.27	0.21
Oxygen (by difference)	3.42	2.89
H/C	1.07	1.13

Corrected Temp, C	Cumulative Volume Distilled % of Sample	Corrected Temp, C	Cumulative Volume Distilled % of Sample
216	IBP	201	IBP
248	10	210	4
253	20	214	17
267	30	225	28
281	40	236	36
299	50	260	47
319	60	310	68
356	70	340	76
406	80	390	86
431	90	430	90
444	95.5	481	94.5
>444	4.5 wt % residue	>481	4.1 wt % residue

(a) Wilsonville SRC pilot plant recycle solvent Sample No. 20232.

(b) Tetralin, J. T. Baker Company Practical Grade.

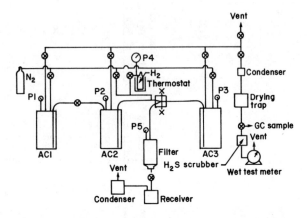

Electric Power Research Institute

Figure 1. Short-residence-time coal liquefaction apparatus (3); Note: *Stirrers in all autoclaves, heater in AC2, and cooling coal in AC3 are not shown. All autoclaves and filters are equipped with furnaces and temperature controls.*

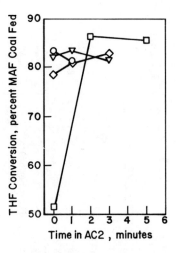

Figure 2. THF soluble conversion vs. time and temperature (3): (□), 413°C; (◇), 427°C; (▽), 441°C; (○), 454°C.

Electric Power Research Institute

SRC Yields. SRC yield was defined as the material which is soluble in the hot filtrate plus the material in the filter cake which is soluble in THF, with a boiling point above 482 C, calculated on an MAF coal and MAF SRC basis. The yields of SRC would normally be smaller than the THF soluble conversions were it not for the presence of the solvent. Components of the solvent may report to the SRC fraction in several way: 1) by reacting with the SRC, this reaction has been studied in some detail by other investigators and found to consist of a reversible reaction of phenols with the SRC product fraction$(\underline{2})$, 2) by polymerizing to higher molecular weight materials which appear in the SRC distillation fraction, and 3) 4-5 weight percent of the starting solvent boiled above 482 C during distillation and was included in the SRC.

Figure 3 is a plot of SRC yield versus time and temperature at 3:1 solvent to coal ratio, uncorrected for the 4-5 percent of solvent which distills in the SRC fraction. SRC yields decrease at 413 C and 427 C and increase at 441 C and probably 454 C as a function of increasing reaction time at temperature. One explanation for these results might be found in the thermal behavior of the solvent. The product of the interaction of SRC with components of the solvent can be reversed with increasing reaction time releasing solvent-molecular weight material. At higher temperatures this is offset by the polymerization of solvent to heavier molecular weight products which distill with the SRC fraction.

There was a slight increase in SRC yield when an overpressure of hydrogen was used.

Quality of SRC. The amount of ash and particularly sulfur in the SRC are as important as the yield. If the SRC is to be used as a clean boiler fuel, the ash content must be quite low and the sulfur content must be low enough to meet the new source standards for SO_2 emissions during combustion. Average analyses for SRC made from both coals are shown in Table 3. As expected, the average SRC produced in the first step of the two-step process, even using a West Kentucky 9/14 coal blend with an unusually low organic sulfur content (Coal Sample 1), results in 1.34 lb SO_2/MM Btu during combustion and would not meet the present new source emission standards. Adding an overpressure of molecular hydrogen did not affect the sulfur level in the SRC when the second step was carried out at 441 C. Higher reaction temperature is effective in lowering the product sulfur content and would probably be required in the second process step.

The sulfur in the first West Kentucky 9/14 coal can be divided into organic (1.16 weight percent) and inorganic (2.61 weight percent) fractions. The maximum percentage of organic sulfur removed in the simulated first step was 37 percent (MAF basis) at 454 C, 1 minute residence time, as shown in Figure 4. Sulfur was removed in increasing amounts with increasing reaction

TABLE III. AVERAGE ANALYSIS OF SHORT RESIDENCE TIME COAL LIQUEFACTION PRODUCTS (3)

Analysis, weight percent of sample	SRC		Light Oil	Filter Cake	Solvent
	1st Coal	2nd Coal			
Moisture	--	--	--	--	--
Ash	0.17	0.15	--	48.4	--
Carbon	84.3	84.9	83.9	42.7	87.75
Hydrogen	5.6	5.75	7.6	2.3	7.7
Nitrogen	1.75	1.7	0.7	1.0	0.73
Sulfur	1.02	1.37	0.19	6.16	0.20
Oxygen (by difference)	6.7	6.2	7.7	3.4	3.77
H/C ratio	.797	.813	1.087	3.4	3.77

Gases, volume percent of sample	Preheater	Reactor
H_2	7.8	11.2
CO_2	34.8	10.9
C_2H_4	0.55	1.3
C_2H_6	3.5	10.4
H_2S	29.0	31.3
CH_4	3.8	23.3
CO	2.25	4.8
Others, C_3^+, by difference	18.83	7.0

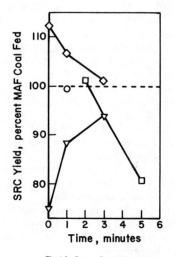

Electric Power Research Institute

Figure 3. SRC yield vs. time and temperature (3): (□), 413°C; (◊), 427°C; (▽), 441°C; (○), 454°C.

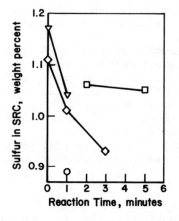

Electric Power Research Institute

Figure 4. Sulfur in SRC vs. time and temperature (3): (□), 413°C; (◊), 427°C; (▽), 441°C; (○), 454°C.

time and at increasing rates with increasing reaction temperature.
There was no effect of increasing solvent to coal ratio on
sulfur removal. This suggests that hydrogen was not being transferred from the Wilsonville recycle solvent to the coal molecule
in order to remove sulfur at the low temperature used in the
first step. The sulfur removal mechanism at this stage probably
only involves thermal removal of nonthiophenic sulfur.

Other Reaction Products. In addition to SRC, gas, light
oil, and a filter cake of unreacted coal and inorganic materials
are produced in the first step of the short residence time coal
liquefaction process. One of the objectives of short residence
time coal liquefaction is to minimize the loss of hydrogen to
gases and light oil.

The average quantity and composition of gases from AC1 and
AC2 are shown in Table 3. The N_2-free gases contained small
quantities of CO_2, H_2S, and other gases which are probably the
result of thermal coal decomposition. Only 2 weight percent of
the as-fed coal goes to gas during the first step of the process.
This amount of gas formation probably cannot be avoided.

Added Tetralin, hydrogen overpressure and increased solvent-
to-coal ratios resulted in no measurable increase in gas yield.
The results are complicated by and corrected for hydrogen added
in the experiments with high-pressure hydrogen.

The H_2S present in the gas in AC2 represents 22 percent of
the sulfur in the coal. The total amount of sulfur released into
the gas phase during the first step of this coal liquefaction
process is 25 percent of the total sulfur in the as-fed coal.
This presumably arises from easily removed organic sulfur and
some of the pyritic sulfur which can be half converted thermally
to H_2S under the reaction conditions.

Individial filter cake compositions vary widely. As conversion increases, sulfur and ash increase while oxygen and hydrogen
and possibly nitrogen concentrations in the filter cakes decrease.
The average filter cake yield is 30 weight percent of the as-fed
coal. The sulfur in the filter cake averaged 49 percent of the
sulfur in the coal feed and is made up of the sulfur remaining
after partial pyrite decomposition and sulfate sulfur.

Hydrogen introduced after removal of unreacted coal and ash
would come in contact with only 25 percent of the sulfur content
of the as-fed coal.

Solvent Composition and Recovery. The solvent was defined
as the product fraction which is soluble in the hot filtrate and
the THF extraction of the filter cake and which boiled between
232 C and 482 C at atmospheric pressure. One of the requirements
of a commercial coal liquefaction process is that a least as much
solvent be created as is used in the process. In addition, the
composition of the solvent must be kept constant if it is to be

used as a hydrogen donor and as a solvating agent for the dissolved coal.

The average solvent recovery for these experiments in the absence of hydrogen or Tetralin was 89 percent, corrected for solvent lost in the residues which remained in the autoclaves and by normalizing mass balances from 95.4 to 100 percent. Of the 11 percent of the solvent unaccounted for directly during the reaction, 5-6 percent is collected in each of the light oil and SRC fractions as discussed earlier. The recovery of solvent, uncorrected for starting solvent which distilled in the light oil and SRC fractions, is shown as a function of reaction time and temperature in Figure 5. There is no correlation with temperature. However, solvent recovery does increase with reaction time and with decreasing solvent to coal ratio.

Hydrogen Transfer. During the liquefaction process, the solvent is presumed to donate hydrogen to the dissolved coal molecules to stabilize them and prevent polymerization reactions which lead to coke. In these experiments, hydrogen transfer was followed by monitoring the elemental analysis of the solvent to see if a change in hydrogen percentage or H/C ratio occurred. Comparison of the average solvent analyses before and after reaction is complicated by the loss of recovered solvent to SRC and light oil fractions which occurs during distillation. The data indicate that there was a net consumption of hydrogen in the presence of high pressure hydrogen. **There is a net production of hydrogen in the absence of high pressure hydrogen.**

The analyses for hydrogen and carbon in the solvent before and after reaction are the same within experimental error. However, if as shown in Tables 2 and 3, hydrogen content in the solvent after reaction has decreased 0.1 weight percent and the H/C ratio has decreased 0.02 on the average, approximately 1.3 percent of the hydrogen in the solvent transfers to coal during liquefaction at 3:1 solvent to coal ratio. This indicates that up to 80 percent of the coal can be converted to THF soluble materials by transferring hydrogen amounting to less than 0.3 weight percent of the coal charge.

The solvent's role in the first step of the process is clearly based as much on dispersing and dissolving the coal molecules resulting from thermal bond breaking as it is on stabilizing the molecules by hydrogen transfer.

Conclusions

In step one, conversion of coal to a THF soluble product is rapid. The THF solubles are unstable in the presence of a coal derived solvent, but in the absence of hydrogen. In step two, the addition of molecular hydrogen to the system or of Tetralin to the solvent to increase hydrogen transfer to the coal increases the THF soluble conversion but does not lower the sulfur

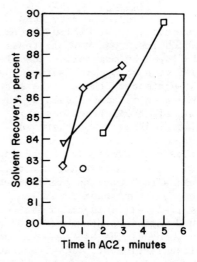

Figure 5.   Solvent recovery vs. time and temperature   (3):   (□),   413°C;   (◇), 427°C; (▽), 441°C; (○), 454°C.

Electric Power Research Institute

content of the SRC. Higher temperatures are required to remove additional sulfur.

SRC yields are greater than 100 percent due to the presence of solvent in the SRC. The average SRC prepared from West Kentucky 9/14 coal blends does not meet new source standards for SO_2 emissions after the first step of the two-step process although all of the inorganic sulfur and an average 12 percent of the organic sulfur are removed.

Hydrogen is produced in the absence of an overpressure of molecular hydrogen and apparently is consumed when hydrogen is present in the system. During step one of the two-step process less than 0.3 weight percent hydrogen is transferred from the solvent to the SRC in the absence of hydrogen. The solvent appears to physically stabilize the coal. Solvent recovery from the first step is approximately 100 percent when corrected for the amounts of starting solvent which are collected in the light oil and SRC fractions.

Acknowledgment

This work was funded by the Clean Liquid and Solid Fuels Group of EPRI, Contract RP-779-5.[3] William C. Rovesti was the Project Manager.

Literature Cited

1. The Nature and Origin of Asphaltenes in Processed Coals, EPRI AF-252, prepared by Mobil Research and Development Corporation, pp 10-18, February, 1976.

2. The Nature and Origin of Asphaltenes in Processed Coals, EPRI AF-480, prepared by Mobil Research and Development Corporation, pp 7-57, 8-24, July, 1977.

3. Short Residence Time Coal Liquefaction, EPRI AF-780, prepared by Battelle's Columbus Laboratories, June, 1978.

RECEIVED June 2, 1980.

Upgrading of Short-Contact-Time Solvent-Refined Coal

R. H. HECK, T. O. MITCHELL, and T. R. STEIN

Mobil Research and Development Corporation, Princeton, NJ 08540

M. J. DABKOWSKI

Mobil Research and Development Corporation, Paulsboro, NJ 08066

In solvent refining, coal is converted to a pyridine-soluble product after very short residence time at reaction conditions. The product formed in this short time is primarily a heavy high-melting point solid that contains most of the organic sulfur, nitrogen and oxygen of the original coal. In the SRC-I and SRC-II processes, this initial product is held at reaction conditions ($\sim$455°C and 13790 kPa) to achieve further defunctionalization and conversion to lighter products. A two-step process in which the coal is dissolved at short residence time, the ash and unreacted coal removed, and the resultant product catalytically hydroprocessed could have significant selectivity and activity advantages over a strictly thermal scheme. In this work, products from short-contact time dissolution were processed over a conventional NiMo/Al$_2$O$_3$ catalyst to determine the feasibility of the catalytic upgrading step in this two-stage liquefaction scheme.

Production of Short-Contact Time SRC

In the solvent-refined coal pilot plant at Wilsonville, Alabama, the coal slurry is heated to reaction temperature in 3-4 minutes residence time in the preheater. The slurry is then held in the dissolver for an additional 40 minutes before it is filtered to obtain specification solvent-refined coal. By bypassing the dissolver and going directly to the filters, samples of short-contact time (SCT) SRC were produced from Illinois #6 (Monterey) and West Kentucky coals.

Table I compares the conditions and results of this operation to those for conventional SRC for Illinois #6 coal. At the short residence time, the coal conversion determined by pyridine solubility is 89% compared to 95% at conventional SRC conditions. The hydrogen consumption and production of light gases are reduced significantly at short residence time, while the SRC yield is increased.

0–8412–0587–6/80/47–139–179$05.00/0

TABLE I

COMPARISON OF REGULAR AND SHORT-CONTACT TIME OPERATIONS

Process Conditions	Regular SRC	SCT-SRC
Temperature, °C		
Preheater Outlet	430	440
Reactor Outlet	457	–
Pressure, kPa	16548	16755
Residence Time, min.	40	3-4
Results, % MAF Coal		
Conversion	95	91
H_2 Consumption	2.9	1.6
SRC Yield	52	76
Gas Make	8.2	1.3

Properties of SCT-SRC

The elemental composition and gradient elution chromatographic (GEC) analysis for two SCT-SRC samples and a regular SRC are compared in Table II.   GEC is a liquid chromatographic technique that was developed by Mobil and is described elsewhere (1).   The hydrogen and nitrogen contents are similar for all three samples.   However, the longer residence time in the dissolver results in lower oxygen and sulfur for the regular SRC sample.   The GEC analyses show that the SCT-SRC contains about 30% of highly polar compounds which are not eluted in this technique.   This compares to only 5% non-eluted material in regular SRC.   This increased fraction of non-eluted materials comes at the expense of polar and eluted asphaltenes.

The distribution of sulfur, nitrogen and oxygen in the GEC fractions is similar for the regular and SCT-SRC from the Illinois #6 coal (Figure 1).   The oxygen content increases from less than 0.8 wt.% in the lowest number, less polar, GEC fraction to almost 10 wt.% in the higher fractions.   The nitrogen content goes through a maximum of about 2.2 wt.% in the resins and eluted asphaltene fractions.   The sulfur is almost evenly distributed across the GEC fractions.

The molecular weights of the SCT-SRC product increase with increasing GEC fraction number and are slightly higher than conventional SRC (Figure 2).   The molecular weight can be correlated with the oxygen content of the GEC fractions for both regular and SCT-SRC (Figure 3).   Data for GEC fractions from H-Coal fuel oil have been included on this figure and show a similar correlation.

TABLE II

ANALYSES OF SHORT—CONTACT TIME AND REGULAR SRC

	SCT West Kentucky SRC 76D3653	SCT Monterey SRC 77D13	Regular Monterey SRC 76D2155
Composition, wt. %			
Hydrogen	6.03	6.14	6.22
Sulfur	0.99	2.19	0.70
Oxygen	5.3	5.0	4.0
Nitrogen	1.99	1.66	1.75
Ash	0.26	0.55	0.085
CCR	44.98	48.40	-
GEC Analysis, wt.%			
No. Fraction Designation			
1 Saturates	0.27	0.04	0.02
2 MNA + DNA Oil	0.21	0.19	1.08
3 PNA Oil	0.21	1.07	1.79
4 PNA Soft Resin	7.54	3.09	4.83
5 Hard Resin	1.53	0.81	2.41
6 Polar Resin	3.53	2.12	4.86
7 Eluted Asphaltenes	12.18	10.67	25.71
8 Polar Asphaltenes	9.94	8.62	15.93
9 " "	2.84	4.62	5.52
10 " "	4.95	6.59	10.97
11 " "	17.58	21.56	17.73
12 " "	9.83	9.11	3.99
13 Non—eluted + Loss	29.39	31.87	5.16
TOTAL	100.00	100.00	100.00

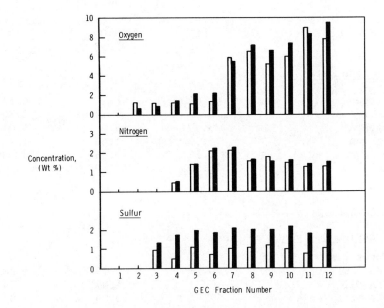

Figure 1.   Elemental composition of short-contact-time and regular Monterey SRC: (□), regular contact time; (■), short contact time.

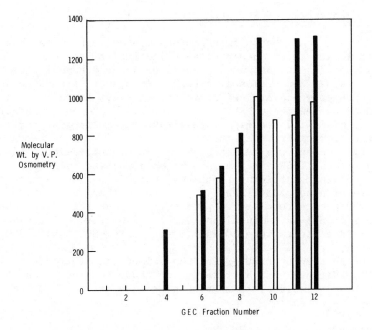

Figure 2. *Molecular weight of GEC cuts from short-contact-time and regular Monterey SRC: (□), regular contact time; (■), short contact time.*

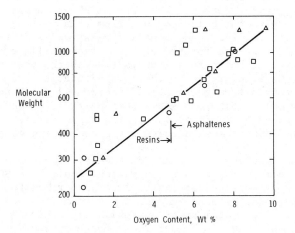

Figure 3. *Molecular weight vs. oxygen content for resins and polar asphaltenes in raw coal liquids: (□), regular SRC; (○), H-coal; (△), SCT–SRC.*

Thermal Processing of SCT-SRC

SCT-SRC could be further processed at temperatures above that
of dissolution to produce a clean solid fuel of reduced sulfur content. Char formation tendency would be lowered by prior removal of
mineral matter and undissolved coal. At higher temperatures, desulfurization would proceed rapidly; light gas formation might be
minimized by keeping the time very short. Hydrogen consumption
would be minimized because aromatic-hydroaromatic equilibria favor
aromatics as temperatures increase.

This concept was examined by thermally processing Illinois #6
Burning Star SCT-SRC in 5 minute batch autoclave runs (2). A synthetic solvent containing about 40% tetralin was used under a H_2
pressure of about 10343 kPa. It was found that temperatures of at
least 480°C were required to produce SRCs containing less than 1%
sulfur from SCT-SRCs containing 1.3 to 2% sulfur. Pyridine-insoluble residue yields were 10-15% (based on SCT-SRC fed) and H_2 con-
sumptions were 0.8 to 1.5%.

For Burning Star SCT-SRCs, the selectivity for desulfurization
is not sufficiently high, relative to those for formation of gases
and insoluble residue, to make this process practicable. No other
SCT-SRCs were tested.

Hydroprocessing of SCT-SRC

The SCT-SRC from the Illinois #6 coal was mixed with regular
SRC recycle solvent to yield a 33% SRC blend. This dilution with
recycle solvent was necessary in order to give sufficient fluidity
for charging to a fixed-bed catalytic reactor. A relatively large
pore, NiMo on alumina catalyst (Harshaw 618X), was employed (Table
III). The results for selected runs from this ten-day study are
given in Table IV. The conditions are relatively severe when compared to those required for hydroprocessing distillate coal liquids
(3). The results show that sulfur is the easiest to remove of the
heteroatoms, while nitrogen is the most difficult. The hydrogen
consumption ranged from about 267 m^3 gas/m^3 oil to greater than
623 m^3 gas/m^3 oil. Figure 4 shows that 65-75% of this hydrogen
winds up in the C_6^+ liquid product, with the remainder going to
make lighter hydrocarbons and heteroatom-containing gases (H_2O,
NH_3, H_2S). Previous studies on regular SRC blends (4) can provide a comparison of the relative ease of upgrading regular and SCT-SRC products. Table V shows charge and product properties for two runs in
which 33% blends of regular and SCT-SRC were hydroprocessed over
the same NiMo/Al_2O_3 catalyst at approximately equivalent conditions.
Although the SCT blend is higher in sulfur and oxygen content and
contains more polar asphaltenes, the products are very similar.
The product from the SCT-SRC charge is actually better in several
respects than that from the regular SRC.

TABLE III

PROPERTIES OF HYDROTREATING CATALYST HARSHAW 618X

Physical Properties

Total Pore Volume, cc/g	0.60
Real Density, g/cc	3.60
Particle Density, g/cc	1.14
Surface Area, m^2/gm	140
Avg. Pore Diameter, Å	171

Chemical Composition, wt.%

Ni	2.7
MoO_3	14.8

TABLE IV

HYDROPROCESSING OF 33% BLEND* OF SHORT-CONTACT TIME SRC
(Pressure 13790 kPa; NiMo Catalyst)

	Feed	Run 1	Run 2	Run 3
Operating Conditions				
Temperature, °C	–	358	385	417
LHSV	–	0.57	0.31	0.57
H_2 Consumption, $m^3 H_2/m^3$ Oil	–	266	479	566
Product Analyses, Wt.%				
Hydrogen	6.5	8.2	9.6	10.0
Sulfur	1.0	0.4	0.1	.07
Nitrogen	1.1	0.8	0.4	.30
Oxygen	4.5	3.0	1.6	.80
CCR	16.8	10.8	5.9	5.4
Yield, Wt.% of Feed				
C_1-C_3	–	0.3	.4	1.9
C_4-C_5	–	0.2	.3	1.2
C_6^+	100.0	98.7	97.7	95.0
H_2	–	-2.0	-3.6	-4.3

*Sixty-seven percent regular SRC recycle solvent.

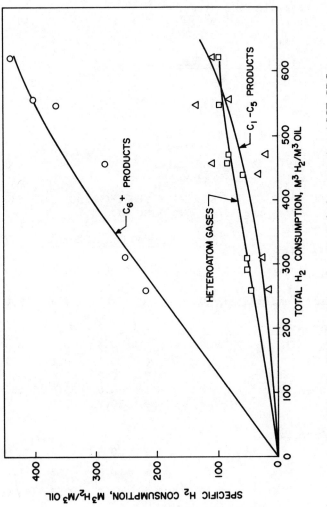

Figure 4.   Specific hydrogen consumption in hydroprocessing of SCT–SRC

TABLE V

HYDROPROCESSING REGULAR AND SCT MONTEREY SRC BLENDS

Processing Conditions	Regular SRC	SCT-SRC
H$_2$ Pressure, kPa	13790	13790
Temperature, °C	414	417
LHSV, V/V/Hr	.50	.57

Liquid Product Properties	Charge	Product	Charge	Product
Gravity, °API	-4.9	12.2	-5.9	12.1
Hydrogen, Wt.%	7.0	9.2	6.5	10.0
Sulfur, Wt.%	0.58	.04	1.01	.07
Nitrogen, Wt.%	1.12	.32	1.14	.30
Oxygen, Wt.%	3.7	1.0	4.5	.8
CCR, Wt.%	17.1	6.2	16.8	5.4
KV (cs @ 100°C)	13.4	1.9	416	2.2
343°C$^-$, Wt.%	48.3	60.7	49.0	66.2
343°C$^+$ Residue, Wt.%				
Cut 1 (Saturates)	0.4	1.4	0.3	2.1
Cut 2-4 (Aromatic Oils)	10.9	24.9	6.0	19.7
Cut 5-7 (Resins/Asphaltenes)	15.9	9.9	8.4	8.2
Cut 8-13 (Polar Asphaltenes)	24.5	3.1	36.3	3.8

The liquid products were distilled to determine the yield and properties of the residual ($343°C^+$) and light liquid ($343°C^-$) products. Table VI shows that Runs 2 and 3 in Table IV resulted in 27 and 34 wt.% conversion of the $343°C^+$ fraction, while the sulfur in this fraction was reduced to 0.25 and 0.18 wt.%, respectively. The distillate and light liquid product ($343°C^-$) are also upgraded in this process. The additional light distillates produced could presumably be recycled to the liquefaction reactor or utilized as low sulfur light distillate fuel.

TABLE VI

ANALYSES OF PRODUCT FRACTIONS FROM HYDROPROCESSING
OF 33% BLEND OF SHORT-CONTACT TIME SRC

	Feed	Run 2	Run 3
IBP-343°C, Wt.%	49.0	62.9	66.2
Hydrogen	7.4	9.7	10.2
Sulfur	1.01	0.012	0.015
Nitrogen	0.6	–	–
Oxygen	4.5	1.1	0.4
$343°C^+$ Bottoms, Wt.%	51.0	37.1	33.8
Hydrogen	5.86	8.25	8.20
Sulfur	1.68	0.25	0.18
Nitrogen	1.66	1.00	0.94
Oxygen	6.00	2.40	1.60

Comparison With Other Liquefaction Processes

Although the work in this study was directed primarily at the upgrading step and was not fully integrated with the dissolution step, the results can be used to estimate the yields for a combination SCT dissolution plus upgrading. (In estimating the overall yields, it was assumed that there was no contribution from conversions of the solvent to formation of gases, naphtha, or residuum; all such products come from upgrading of the SCT-SRC.) The yield of these products was added to the yield from the SCT dissolution step at Wilsonville to obtain an estimate of the overall process. No attempt was made to optimize the hydrotreating step or to integrate the process with respect to solvent recycle between the upgrading and dissolution steps.

Table VII shows the estimated yields from an SCT-SRC plus hydrotreating scheme along with published yields from SRC-I (5), SRC-II (6) and H-coal Syncrude (7) processes. The yields for the SCT dissolution operation at Wilsonville (8) are also included in this table. All yields are based on Illinois #6 coal except those for the SRC-II process which is based on a West Kentucky bituminous coal. All yields were adjusted to an MAF basis based on coal fed to the liquefaction reactor.

Comparison of coal liquefaction process yields is difficult in that each scheme separates the products into different boiling ranges when reporting yields. In addition, some processes, such as SRC-II or H-Coal Syncrude, only report distillate and lighter products; the residual materials being used for hydrogen production, while in other processes, such as SRC-I, the residual material is the primary product and hydrogen must be produced from raw or unreacted coal which is separated from the process. With these reservations in mind, the processes were compared on a net liquid yield from the liquefaction step, exclusive of any consideration for hydrogen production.

Table VII shows that the SCT-SRC plus upgrading yields significantly less gas and more liquid (residual material included) than the other processes. The hydrogen consumption in the two-step SCT process is higher than for the SRC-I process; however, it is still lower than for the SRC-II process and significantly lower than for the H-Coal Syncrude operation.

Table VIII shows the yield and approximate elemental composition of the composite liquid product from these processes. The SCT-SRC and SRC-I residual products were assumed to be liquids for this comparison. SCT-SRC plus upgrading gives a liquid yield almost equivalent to the SRC-I process, but the liquid is of significantly higher quality: 2.4% higher in hydrogen and 2884 kJ/kg higher in heating value. The H-Coal Syncrude process yields a somewhat higher value liquid in terms of hydrogen content. However, this liquid is higher in sulfur and oxygen content, which tends to reduce its heating value. The total liquid yield from the H-Coal process is again significantly lower than from the two-step SCT process.

Although the comparisons are by no means exact, a process based on short-contact time dissolution and catalytic upgrading would appear to have potential for significantly higher yields of high quality liquids from coal.

Conclusion

Illinois #6 and West Kentucky coals are dissolved at short residence times in the SRC process. When compared to regular SRC, short residence time operation results in a decreased production of light liquids and gases and less hydrogen consumption. The solid product from short contact time is more viscous and higher in sulfur and oxygen content than conventional SRC. Although the short-contact time product is a lower quality SRC, the differences between regular and short-contact time SRC disappear rapidly with catalytic hydroprocessing. This processing yields an upgraded, low sulfur boiler fuel and produces additional solvent for recycle or use as a light distillate fuel.

TABLE VII

ESTIMATED YIELDS FROM LIQUEFACTION SCHEMES

	SRC-II	SRC-I	SCT-SRC	SCT-SRC HDT	H-Coal Syncrude
Yields % MAF					
Light Gases	18.4	8.2	1.3	4.0	11.8
Hydrogen	-5.2	-2.9	-1.6	-4.9	-6.6
Distillate Products					
Approximate Boiling Range, °C	C_5-482	C_6-371	C_6-371	C_5-343	C_4-524
Yield, % MAF	48.5	27.4	10.1	31.1	34.7
Residual Products					
Yield, % MAF	22.4	52.4	75.8	48.6	47.5
Unreacted Coal	4.1	5.0	9.2	9.2	-

TABLE VIII

ESTIMATED YIELD AND COMPOSITION OF LIQUID PRODUCTS FROM
LIQUEFACTION PROCESSES

	SRC-II	SRC-I	SCT	SCT-SRC HDT	H-Coal Syncrude
Liquid Yield, Wt.% MAF	48.5	79.8	85.9	79.7	34.7
Approximate Boiling Range	38-482°C	C_5^+	C_5^+	C_5^+	C_4-524°C
Elemental Analyses, Wt.%					
Hydrogen	8.8	6.3	6.3	8.7	10.5
Sulfur	0.3	0.6	2.0	0.2	0.8
Nitrogen	0.8	1.3	1.5	0.7	0.4
Oxygen	3.9	4.1	4.9	1.6	2.4
Estimated Heating Value (kJ/kg)	40077	38053	37146	40938	41915

A comparison of potential yields from a combined SCT dissolution plus upgrading scheme with yields from other liquefaction schemes shows that the SCT scheme has potential to give significantly lower gas and higher liquid yields. This results in a more efficient utilization of hydrogen in the liquefaction process.

Acknowledgement

This work was conducted under Electric Power Research Institute (EPRI) Contract No. RP-361-2 which is jointly funded by EPRI and Mobil Research and Development Corporation. Figures 1, 2, and 3 in this chapter appeared in EPRI Report AF-873.

Literature Cited

1. Bendoraitis, J.G.; Cabal, A.V.; Callen, R.B.; Stein, T.R.; Voltz, S.E. Phase I Report, EPRI Contract No. RP 361-1, Mobil Research and Development Corporation, January 1976.

2. Dickert, J.J.; Farcasiu, M.; Mitchell, T.O.; Whitehurst, D.D. "The Nature and Origin of Asphaltenes in Processed Coals", Final Report to EPRI under contract RP 410-1, in press, 1979.

3. Bendoraitis, J.G.; Cabal, A.V.; Callen, R.B.; Dabkowski, M.J.; Heck, R.H.; Ireland, H.R.; Simpson, C.A.; Stein, T.R. Annual Report, EPRI Contract No. AF-444 (RP 361-2), Mobil Research and Development Corporation, October 1977.

4. Stein, T.R.; Cabal, A.V.; Callen, R.B.; Dabkowski, M.J.; Heck, R.H.; Simpson, C.A.; Shih, S.S. Annual Report, EPRI Contract No. AF-873 (RD 361-2), Mobil Research and Development Corporation, December 1978.

5. "Solvent Refined Coal Pilot Plant, Wilsonville, Alabama", Technical Report No. 8, Catalytic Incorporated to Southern Services, April 30, 1976.

6. Schmid, B.K. and Jackson, D.M. The SRC-II Process, Third Annual International Conference on Coal Gasification and Liquefaction, University of Pittsburgh, August 3-5, 1976.

7. "H-Coal Process Passes PDU Test," Oil and Gas Journal, August 30, 1976, 52-53.

8. "Operation of Solvent Refined Coal Pilot Plant at Wilsonville, Alabama," Annual Technical Progress Report, January-December 1976.

RECEIVED June 12, 1980.

Processing Short-Contact-Time Coal Liquefaction Products

CONRAD J. KULIK, HOWARD E. LEBOWITZ, and WILLIAM C. ROVESTI

Electric Power Research Institute, 3412 Hillview Avenue, Palo Alto, CA 94303

A considerable effort has been expended in the past few years by many researchers in attempts to better understand the mechanism by which coal is liquefied. From this work has emerged the concept of short residence time coal liquefaction which promises potential process advantages, small reactor, minimum hydrogen flow, and the efficient utilization of hydrogen for a particular product slate.

Work done for EPRI by Mobil Research (1) and Battelle (2) demonstrated that coal could be liquefied at these relatively short reaction times. This work, however, was limited, and indicated some very apparent process disadvantages:

- process was out-of-solvent balance,
- a viscous reactor effluent was produced resulting in poor filterability,
- vacuum still bottoms had high viscosity,
- the product was thermally sensitive.

In order to overcome these problems, the flow schemes as shown in Figures 1 and 2 were developed. These incorporate the use of Kerr-McGee Corporation's Critical Solvent Deashing and Fractionation Process (CSD) for recovery of the SRC. The Kerr-McGee Process adds extra flexibility since this process can recover heavy solvent for recycle, which is not recoverable by vacuum distillation. EPRI contracted with Conoco Coal Development Company (CCDC) and Kerr-McGee Corporation in 1977-1978 to test these process concepts on continuous bench-scale units. A complementary effort would be made at the Wilsonville Pilot Plant under joint sponsorship by EPRI, DOE, and Kerr-McGee Corporation. This paper presents some of the initial findings.

Experimental

CCDC built a continuous short residence time coal liquefaction unit with throughput of about 4.5 kg/hr of coal. The SRC unit consisted of a short residence time reactor constructed from 53.3 m of high pressure tubing having an ID of 0.516 or

0–8412–0587–6/80/47–139–193$05.00/0

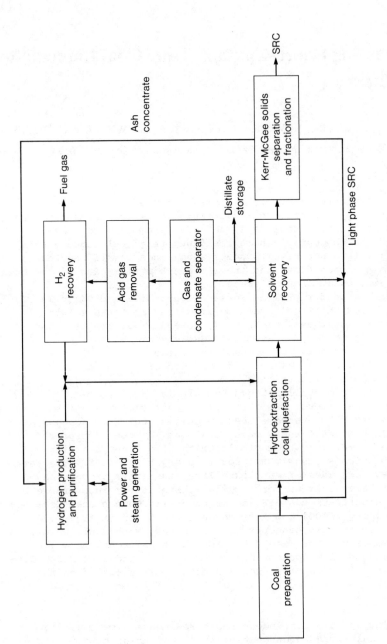

Figure 1.   Coal hydroextraction with light-phase SRC recycle process

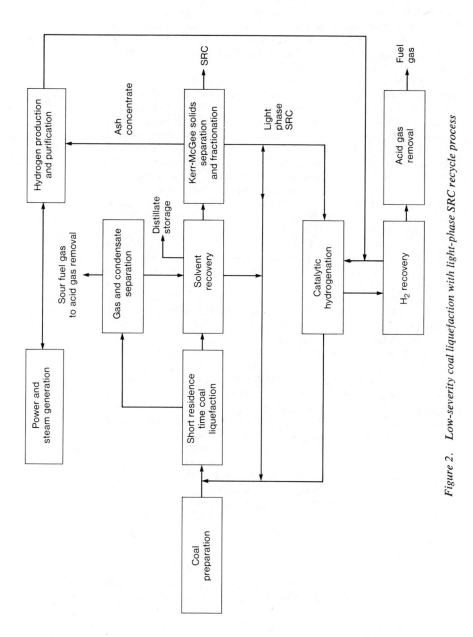

Figure 2. Low-severity coal liquefaction with light-phase SRC recycle process

0.704 cm. The coil was heated by a radiant furnace with four
individually controlled heating zones. The furnace was controlled to simulate a linear heat-up profile. The bench-scale
coil was operated in the laminar flow region, where coking can
be a problem. On a commercial scale, this furnace would operate
in a highly turbulent mode. During the early phases of work at
CCDC, work centered around operation of the continuous bench-
scale SRT unit with distillable solvents.

Paralleling the work at CCDC were the critical solvent
deashing and fractionation studies done on a continuous bench-
scale unit at Kerr-McGee Technical Center, Oklahoma City, Oklahoma, Figure 3. The Kerr-McGee Critical Solvent Deashing and
Fractionation Process has been previously discussed (3).

In work prior to this program, Kerr-McGee demonstrated that
extremely rapid solids separation (deashing) on the order of 30
to 60 times faster than a conventional deasphalting unit and
high deashing efficiencies producing less than 0.1 wt% ash on
SRC product could be achieved. In addition, it has been demonstrated that the SRC could be fractionated into multiple residual fractions.

The work involved the integration of the SRC operations at
Kerr-McGee with those at CCDC where the concept of recycling
certain residual fractions back to liquefaction would be
tested. This program involved repeated product shipments
between the respective laboratories. The data presented in this
paper will focus on the work done in this latter phase of the
program.

In addition to continuous bench-scale work, CCDC carried
out a rather extensive laboratory program involving the use of
the microautoclave reactor. The program developed tests to
compare the activities of different solvents. These tests
quickly evaluated a solvent so that the performance under coal
liquefaction conditions could be predicted. The tests are now
used at the Wilsonville SRC Pilot Plant as a means of determining when stable operation has been achieved.

The microautoclave solvent activity tests measure coal
conversion in a small batch reactor under carefully controlled
conditions. The tests are described as Kinetic, Equilibrium and
SRT. The Kinetic and Equilibrium Tests measure coal conversion
to tetrahydrofuran solubles at conditions where conversion
should be monotonically related to hydrogen transfer. The
Kinetic Test is performed at 399°C for 10 minutes at an 8 to 1
solvent to coal ratio. The combination of high solvent ratio
and low time provide a measure of performance at essentially
constant solvent composition. The measured conversion is thus
related to the rate of hydrogen donation from solvent of roughly
a single composition. In contrast, the Equilibrium Test is
performed at 399°C for 30 minutes at a 2 to 1 solvent to coal
ratio. At these conditions, hydrogen donors can be substantially depleted. Thus performance is related to hydrogen donor

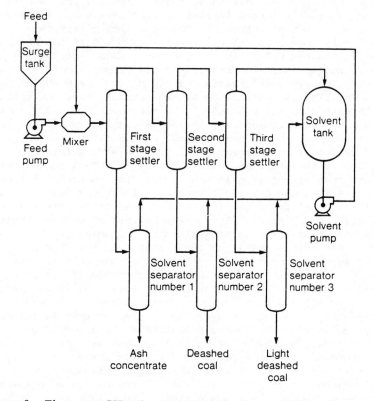

Figure 3. Three-stage CSD pilot plant; blockflow diagram—Wilsonville program.

concentration. The SRT Test, performed at 427°C for 5 minutes
at a 2 to 1 solvent to coal ratio, simulates performance at
short residence time coal liquefaction conditions.

Discussion of Results

   Autoclave Results - Solvent Activity Test. The initial
microautoclave work was done with tetralin and methylnaphtha-
lene, using Indiana V bituminous coal (Table I). Base line data
is shown in Figure 4. All three tests, Kinetic, SRT, and Equilibrium, show an increase in coal conversion with an increase in
the concentration of tetralin. The Equilibrium Test shows the
highest coal conversion of approximately 86 wt% of the MAF coal
(based on the solubility in the tetrahydrofuran) at the 50%
tetralin concentration. The Kinetic Test shows lower coal conversion. The hydrogen transferred to the coal from the tetralin
in the Equilibrium Test at the 50 wt% tetralin feed concentration is approximately 0.5 wt% of the MAF coal. In the Kinetic
Test 50 wt% tetralin feed concentration results in a much
smaller transfer at the short reaction time of 10 minutes.
   Microautoclave data was also obtained with Wilsonville
Batch I solvent utilizing Indiana V coal. Batch I solvent was
obtained from Wilsonville in mid-1977. Other batches of recycle
solvent were received later. Batch I solvent had inspections
most like the Allied 24CA Creosote Oil used for start-up at the
Wilsonville Pilot Plant. Succeeding batches of solvent received
by CCDC showed substantial differences, presumably due to equil-
ibration at various operating conditions. As the Wilsonville
solvent aged and became more coal derived, the solvent aromaticity decreased with an increase in such compounds as indan and
related homologs. The decrease in aromaticity has also been
verified by NMR. A later solvent (Batch III) also showed an
increase in phenolic and a decrease in phenanthrene (anthracene)
and hydrogenated phenanthrene (anthracene) type compounds.
   The hydrogen content of Batch I solvent was varied by catalytic hydrogenation in a fixed bed, trickle phase, adiabatic
reactor at various severities. American Cyanamid HDS-3A, a
nickel-molybdenum catalyst, was used. Reactor conditions were
varied from 8.4 to 11.1 MPa and from .5 to 2 LHSV at a reactor
temperature of 371°C and a hydrogen to feed ratio of .14 m^3
hydrogen per .45 kg of feed. At these hydrogenation conditions,
hydrogenated Batch I solvent was produced with various hydrogen
contents. The optimum coal conversion under SRT Test conditions
was obtained with Batch I solvent to which 1 wt% hydrogen was
added. With solvent hydrogen contents above 9 wt%, the coal
conversion slowly decreased, indicating that even though the
hydrogen content of the solvent was increased, the additional
chemical hydrogen was not being made available as hydrogen
donors at this reaction severity.

TABLE I

ANALYSES OF COALS

	MOISTURE	PROXIMATE (AS RECEIVED) WT %		
		VOLATILE MATTER	FIXED **CARBON	ASH
INDIANA V. (OLD BEN)	4.35	38.22	47.34	10.09
ILLINOIS 6 (BURNING STAR)	3.61	37.87	47.92	10.60

	ULTIMATE (MOISTURE FREE) WT%				SULFUR FORMS				ASH
	H	C	N	O	TOTAL	PYRITE	SULFATE	ORGANIC	
INDIANA V	4.57	69.22	1.36	10.68	3.62	1.07	0.65	1.90	10.55
ILLINOIS 6	4.73	70.05	1.39	9.60	3.23	1.13	0.38	1.72	11.00

	ASH ANALYSIS, WT%										
	Na_2O	K_2O	CaO	MgO	Fe_2O_3	TiO_2	P_2O_5	SiO_2	Al_2O_3	SO_3	OTHER
INDIANA V	0.44	2.14	3.88	0.75	23.06	1.22	0.21	45.22	20.34	0.26	2.48
ILLINOIS 6	0.61	1.84	6.42	0.88	15.68	1.00	0.07	45.18	17.31	7.06	3.95

	WT % ON TYLER SCREEN (WET)				
	48	100	200	325	-325
INDIANA V	0.0	0.6	4.7	32.2	62.5
ILLINOIS 6	2.6	12.0	23.3	15.7	46.4

Figure 5 shows that the solvent hydrogen donor content
plays a significant role in liquefaction performance at short
reaction times. A coal conversion maxima is reached at approximately 4 minutes after which measured coal conversion decreases
due to regressive reactions (reconversion of the THF Solubles to
THF Insolubles). Understandably, other coal conversion per-
formances can be expected with different solvent qualities and
other reaction temperatures. Additional work has shown that not
only the total hydrogen donor content of the solvent is important but also the activity of the donors present, i.e., heavier
molecular weight hydroaromatics such as hydrophenanthrene donate
hydrogen more rapidly than tetralin. The effect of solvency
also is a factor and it appears the "heavier" the solvent (i.e.,
higher boiling point or molecular weight) the better the performance. The interrelationship of the amount and type of hydrogen
donors along with the solvency effect at a specific set of reaction conditions appear to be dictating liquefaction performance
particularly at short reaction times.

The superiority of "heavy" solvents appears to refute the
proposition that the rate controlling step in coal conversion is
the pyrolysis of the coal and that given a sufficient concentration of donors, the rate of hydrogen donation would not be
limiting (4). Comparing the performance of natural solvents to
tetralin, the factor that appears to be limiting conversions is
the hydrogen donation rate of tetralin, at least in the early
stages of coal dissolution. Hydrogen donors contained by coal
derived solvent reacted more rapidly than tetralin. Work by
Whitehurst (1) has also shown the same phenomenon.

The most dramatic discovery in the microautoclave study was
the enhancement of coal conversion with Light SRC addition (see
Table II). Success with the high boiling distillable solvents
encouraged experimentations with Light SRC. Light SRC is
obtained by fractionating SRC in the critical solvent. A
50% (wt) blend of Light SRC and 256 x 535°C Batch III solvent
was tested on the microautoclave as shown in Table II. The 50%
blend performed well in the Kinetic Test and rather poorly in
the Equilibrium Test as compared to the distillate base. This
presumably is indicative of a low concentration of highly active
donors. When 7 MPa cold gaseous hydrogen was added to the
microautoclave, and the Kinetic, Equilibrium Tests reperformed,
a rather remarkable phenomenon occurred. The results of the
Kinetic Test remained unchanged, but in the Equilibrium Test,
the addition of gaseous hydrogen caused a higher coal conversion. It could be surmised that the gaseous hydrogen reacted
with the coal liquefaction media even at the low temperature of
399°C, where one would have expected the addition of a catalyst
to be required for aromatic hydrogenation. To further understand this phenomenon, a base run was made with Batch III solvent to which gaseous hydrogen was added. No change was
apparent in the Equilibrium Test result. The Light SRC is thus

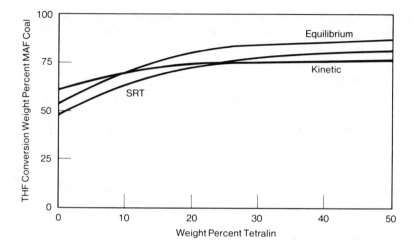

Figure 4. Microautoclave tests, Indiana V coal. Tetralin–methyl naphthalene mixtures (conversion vs. percent Tetralin content)

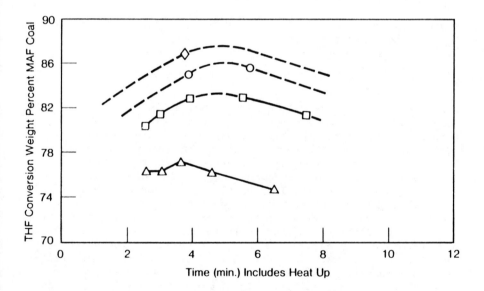

Figure 5. Microautoclave tests, Indiana V Coal (conversion at 440°C vs. time). Batch I solvent: (○), 8.7% hydrogen solvent 3/1 S/C; (◇), 8.9% hydrogen solvent 3/1 S/C; (□), 8.0% hydrogen solvent 3/1 S/C; (△), 8.0% hydrogen solvent 2/1 S/C.

a curious material having few donors of its own, but promoting the reaction of other donors and the reaction with gaseous hydrogen.

TABLE II

MICROAUTOCLAVE DATA-LIGHT PHASE SRC ADDITION-INDIANA V COAL

	THF CONVERSION		
	KINETIC	SRT	EQUILIBRIUM
WHOLE WILSONVILLE BATCH III	76.5	71.5	74.4
WHOLE WILSONVILLE BATCH III W/7 MPa OF COLD H_2	-	-	74.4
50WT% LIGHT PHASE SRC 50WT% 256 x 535°C BATCH III	87.5	72.5	65.5
50WT% LIGHT PHASE SRC 50WT% 256 x 535°C BATCH III W/7 MPa OF COLD H_2	87.5	75.7	86.6

Continuous Bench-Scale Experimentation. With encouraging results obtained from microautoclave tests, experimentation emphasis moved to the bench-scale unit. Here the concept of adding Light SRC to the recycle solvent on a continuous basis was tested. Earlier work (5) performed on short contact time coal liquefaction showed Indiana V coal to be out-of-solvent balance. Also the operability of the continuous bench-scale SRT unit was highly dependent upon the quality of the solvent. Short residence time vacuum bottoms were processed in the Critical Solvent Deashing and Fractionation Unit to allow recovery of higher boiling solvent that would not normally be recovered by distillation. It was postulated that recycle of this material back to liquefaction would help close the solvent balance and improve SRT unit operability. In light of the qualities of the Light SRC found in microautoclave tests, the initial phase of the continuous work was expanded toward testing the concept of Light SRC recycle in the conventional SRC-I mode with an Illinois No. 6 Coal from Burning Star No. 2 mine. Analyses are given in Table I. Burning Star coal was chosen since it has low solvent range yields in ordinary SRC operations.

The work was done on the continuous bench-scale hydro-extraction unit at CCDC which was previously described (6). The distillate solvent and Light SRC for this program were obtained from the Wilsonville Pilot Plant. The processing history by which the recycle solvent was produced at the Wilsonville Pilot Plant was somewhat different from the processing conditions planned for the solvent on the bench scale unit. It was feared that the Wilsonville solvent may have contained residual hydrogen donors that would not be available at the bench unit operating conditions. The solvent was therefore recycled for each series of runs. Each series of runs constituted ~ 60 hours of operations after which the distillate solvent was recovered and recycled to the next series. A base case run was made in each series to identify changes in the distillate process solvent which could be attributed to depleting residual hydrogen donors as the solvent was being recycled from series to series. The data showed, however, no appreciable change in quality as the solvent was recycled. The Light SRC, obtained from the Wilsonville Pilot Plant, was used on a once-through basis except for Runs 4 and 5 where internally recovered Light SRC was used.

Figure 6 shows that as the concentration of the Light SRC was increased the yield of distillable recycle solvent (+206 x 535°C) also increased. With 30 wt% of Light SRC in the total solvent the net yield of recycle solvent was zero, e.g., the process was in solvent balance. Without the addition of Light SRC, in Run 1A, there was a net recycle solvent deficit of approximately 15% of the MAF coal. It should be noted that the plotted distillate yields are only for the liquefaction unit with vacuum distillation. If yields are obtained around the liquefaction and CSD unit the distillate yields are appreciably higher due to the recovery of heavy distillate on the CSD unit that would not normally be recovered by distillation. These values are footnoted in Figure 6. The Light SRC addition had then demonstrated a very dramatic improvement in liquefaction performance even at these mild operating conditions. From each of the runs with Light SRC addition, Kerr-McGee recovered on the CSD Bench-Scale Unit Light SRC approximately equivalent to the amount of Light SRC required to sustain Light SRC recycle.

A second series of runs was made that investigated the effect of liquefaction temperature on yield performance with 30 wt% Light SRC addition. Interestingly, the lower temperatures resulted in high SRC yields with low gas and water yields and sufficient recycle solvent to sustain recycle. The hydrogen consumption was low as expected in the order of 2 wt% on MAF coal. From this data it appeared that these mild operating conditions were conducive in producing SRC with very efficient hydrogen utilization. Further work was done at these conditions, as shown in Figure 7, to determine the effect of recycling Light SRC at these mild operating conditions, 418°C and 8.3 MPa. Light SRC for recycle was recovered by CSD fractionation from some of the runs previously described.

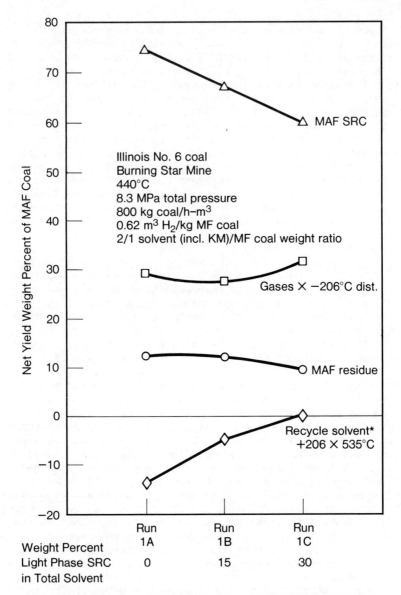

*The plotted yields are for liquefaction alone. The combined yields including CSD are as follows:

Run	Yield 206 × 535°C
1B	0.3
1C	10.2

Figure 6.   Hydroextraction yields

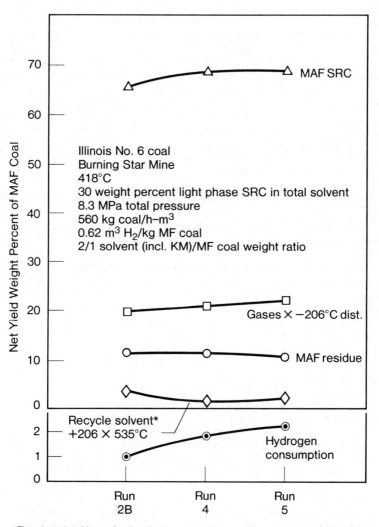

Figure 7.   Recycle hydroextraction yields

Data from Runs 4 and 5 (Figure 7) show that with repeated Light SRC recycle the product yields remained essentially constant with the exception that the hydrogen consumption increased slightly. From each of these runs Kerr-McGee recovered on the CSD Bench-Scale Unit Light SRC approximately equivalent to the amount of Light SRC required to sustain recycle of the Light SRC. The composition of the recovered Light SRC in Runs 4 and 5 was nearly identical in composition to the Light SRC initially added to Run 2B. In addition, viscosity measurements made on +535°C vacuum bottoms made with Light SRC addition showed a great reduction in viscosity as opposed to vacuum bottoms made under similar conditions without Light SRC addition. The addition of Light SRC should improve the operability of the vacuum bottoms handling particularly with short residence time products that are high in preasphaltenes.

Continuous Short Residence Time Experimentation. After determining the effect of Light SRC addition to a conventional SRC-I operation, experimentation moved to determining the effect of Light SRC addition on short residence time coal liquefaction performance.

Earlier short residence time work (5) had shown the liquefaction of Indiana V coal to be out-of-solvent balance and that the operability of the SRT unit was particularly sensitive to the quality of recycle solvent. Batch VI solvent to be used in this third phase of the program was the latest in the series of solvents received by CCDC from Wilsonville. This Batch VI solvent was of a lower quality than Batch I solvent which was operable on the SRT unit in the donor mode but very similar in quality to Batch III. Attempts to run Batch III solvent on the continuous Bench-Scale SRT unit were unsuccessful in the hydrogen donor mode. Gaseous hydrogen addition, at elevated pressure, was required. Attempts to improve the later series of solvents by catalytic hydrogenation proved unsuccessful. This later phase of the program had the objective of determining whether the addition of Light SRC would improve not only SRT unit operability in the hydrogen donor mode but also help to close the solvent balance.

Table III shows the results of operating the SRT unit in the hydrogen donor mode (catalytically hydrogenated solvent) with and without the addition of Light SRC to the distillate solvent. Batch I solvent was used in Run 9. A blend of Batch VI solvent and Light SRC, 70/30 weight ratio, were catalytically hydrogenated as the feed to Runs 1 and 3. The hydrogen donor capability of the solvents were measured by the Equilibrium microautoclave tests. These bench-scale SRT results are rather extraordinary in respect to increased distillate yields and improvement in unit operability with addition of Light SRC. In Table III the integrated yields refer to the combination of liquefaction, CSD, and catalytic hydrogenation of the solvent.

TABLE III

SRT LIQUEFACTION-HYDROGEN DONOR MODE WITHOUT GASEOUS HYDROGEN

SRT RUN NO.	MAX. PROCESS TEMP. °C	OPERATING PRESSURE MPa	% LIGHT SRC	RESIDENCE TIME ABOVE 316°C MIN.	CONVERSION (CRESOL)	535+°C SRC	WT% MAF COAL	
							C_6x206°C	206x535°C
Liquefaction Unit Alone								
9 (1)	440	3.5	0	1.20	78.4[3]	76.4[3]	-3.0	-3.6
1 (2)	441	3.5	30	1.54	80.6	70.0	0.8	1.0
3 (2)	441	3.5	30	0.55	83.6	75.1	2.0	-0.8
Integrated Performance								
1(2)	441	3.5	30	1.54	80.6	55.2	6.8	10.4
3(2)	441	3.5	30	0.55	83.4	60.0	5.5	7.4

(1) Hydrogenated Wilsonville Solvent
(2) Hydrogenated Blend
(3) Conversion Measured in THF rather than CRESOL

To evaluate the effect of adding gaseous hydrogen directly to the SRT unit without externally hydrogenating the solvent, one run was made with the addition of Light SRC. Table IV shows the effects of Light SRC addition and again increased distillate yields are noted. The indication is that the process is in solvent balance. Further work is required on an integrated basis recycling both Light SRC and distillate solvent to further substantiate these initial findings.

Kerr-McGee CSD Performance as Related to Product Quality. As previously mentioned, the final phase of the program involved cyclic shipments between CCDC to Kerr-McGee Corporation. Vacuum bottoms produced via conventional SRC-I or SRT liquefaction modes were sent to Kerr-McGee for critical solvent deashing and fractionation. In some instances the recovered Light SRC was sent back to CCDC for recycle. Kerr-McGee attempted to recover an amount of Light SRC required to maintain recycle. In most instances an equivalent amount of Light SRC was recovered to sustain recycle in both the conventional SRC and SRT modes. Approximately 30% of the MAF coal was lost to the rejected Kerr-McGee phase, ash concentrate, as SRC, and it appeared that the amount of SRC lost to the ash concentrate as a percent of the MAF coal was essentially constant and independent of a wide range of liquefaction severities. If the CSD performance was expressed as a fraction of the SRC produced, the CSD performance was highly dependent upon the quality of the SRC. Higher pre-asphaltene content corresponded to lower SRC recovery. At the extreme limit, low residence time, no gaseous hydrogen addition and high temperature, the SRC product is of a very poor quality, high preasphaltene content (Table V). Here, considerably more than the 30 wt% of MAF coal was left in ash concentrate.

It is interesting to note that the comparison run, made at higher liquefaction severity, produced a comparable preasphal-tene content. But upon further examination of these products by Mobil's SESC analyses, a noticeable difference between the pro-ducts was observed. The lower severity product showed a higher content of the higher SESC fractions. Unfortunately, the work done between CCDC and Kerr-McGee was performed in a blocked-out fashion which necessitated the reheating of various products at either Kerr-McGee or CCDC which can result in thermal degrada-tion of the products.

In Wilsonville Runs 143 and 147, thermal degradation of the coal-derived products greatly affected the SRC recovery on the Kerr-McGee CSD Unit. Both runs were made at identically the same operating conditions, except in Run 143, where presumably catalytically active solids were allowed to accumulate in the liquefaction reactor, whereas in Run 147 the solids were removed. The product yields exiting the reactor for both runs were very similar; however, the thermal sensitivity of the

TABLE IV

SRT LIQUEFACTION—DIRECT HYDROGENATION WITH GASEOUS HYDROGEN

SRT RUN NO	MAX. PROCESS TEMP. °C	OPERATING(1) PRESSURE MPA	% LIGHT SRC	RESIDENCE TIME ABOVE 316°C MIN.	CONVERSION (CRESOL)	WT% MAF COAL		
						535+°C SRC	C_6x206°C	206x535°C
Liquefaction Unit Alone								
22	440	13.9	0	3.0	81.8[2]	76.6[2]	2.0	-6.0
2	441	13.9	30	3.3	87.8	75.0	5.1	-2.9
Integrated Performance								
2	441	13.9	30	3.3	87.8	66.3	5.1	6.6

(1) Hydrogen treat rate 0.11m^3/kg
(2) Conversion measure in THF rather than CRESOL

TABLE V

PRODUCT RECOVERY OF SHORT CONTACT TIME FEEDS

LIQUEFACTION CONDITIONS	SRT-4	SRT-2
EXIT TEMPERATURE, °C	454	441
TIME ABOVE 316°C, MIN.	0.6	3.3
HYDROGEN GAS, MPa	NONE	137
NET CSD SOLIDS FREE FEED ANALYSIS*, WT%		
BENZENE SOLUBLE	27	25
BENZENE INSOLUBLE	73	75
NET RECOVERY IN CSD, WT% OF SOLIDS FREE FEED	30	58

*   Excluding the amount of Light SRC required to sustain recycle.

TABLE VI

EFFECT OF PRESSURE AT WHICH SRC WAS PRODUCED ON CSD RECOVERY

WILSONVILLE COMMON OPERATING CONDITIONS

KENTUCKY 6/11 COAL

800 kg/hm^3

440°C

RUN NO.	150	151
SRC REACTOR PRESSURE (MPa)	11.7	14.5
CONVERSION (MAF COAL %)	94	94
SRC YIELD (MAF COAL %)	65-67	59-61
% SRC RECOVERY IN CSD (OPTIMIZED)	78%	85%

products were vastly different. Products produced from Run 147 degraded most readily to "post-asphaltenes." This resultant degradation lowered the CSD SRC recovery. Additional work at Wilsonville showed that the CSD performance is linked directly to the quality of SRC produced (Table VI).

A question then arises as to whether the CSD recovery is being limited by the preasphaltene content produced from direct products of coal liquefaction or whether by low liquefaction severity a more thermally sensitive product is produced resulting in retrogressive reactions of liquefaction products to "post-asphaltenes." There is some indication that "virgin" preasphaltenes, primary products of coal dissolution, are more easily recovered via CSD as shown in Table VII; however, "post-asphaltenes" made from thermal regressive reactions are not. The species are inseparable by ordinary analytical measures. Further work is being done to more clearly understand the role of regressive reactions in low severity liquefaction. In addition, recent work has resulted in techniques for obtaining high SRC recoveries from less desirable feedstocks.

TABLE VII

CSD RECOVERY OF A SHORT CONTACT FEED

● FEED PRODUCED AT WILSONVILLE (INDIANA V COAL)

● CONTAINED ABOUT 1/3 DISTILLATE PRODUCTS (MOSTLY OILS)

● CSD FEED ANALYSIS, WT% OF SOLIDS FREE FEED

OIL (includes Distillate)	30.3
ASPHALTENE	35.7
PREASPHALTENE	33.9

● CSD PRODUCT RECOVERY, WT% OF FEED COMPONENT

OIL	94.4
ASPHALTENE	88.6
PREASPHALTENE	74.6

Conclusions. The quality of liquefaction solvent is an extremely important factor in liquefying coal at conventional or short residence time liquefaction conditions. The ability to alter the quality of this solvent by recycle of certain SRC fractions has made a marked improvement in the liquefaction performance over a wide range of liquefaction severities. The implication of these findings as to a finalized overall process scheme has yet to be determined; however, this work supports the underlying process concept of being able to efficiently utilize hydrogen to produce a particular product slate. Further work is needed on an integrated basis to substantiate these initial findings.

Literature Cited

1) D. D. Whitehurst, M. Farcasiu, and T. O. Mitchell, "The Nature and Origin of Asphaltenes in Processed Coals," EPRI Report AF480, Annual Report, RP410, July 1977.

2) J. R. Longanbach, J. R. Droege, and S. P. Chauhan, "Short Residence Time Coal Liquefaction," EPRI Report AF780, Final Report, RP779-5, June 1978.

3) R. M. Adams, A. H. Knebel, and D. E. Rhodes, "Critical Solvent Deashing of Liquefied Coal," American Institute of Chemical Engineers, Miami, Florida, November 15, 1978.

4) G. P. Curran, R. T. Struck, and E. Gorin, Ind. and Eng. Chemistry Proc. Des. and Dev. 6, No. 2, 166 (1967).

5) J. A. Kleinpeter, F. P. Burke, P. J. Dudt, and D. C. Jones, "Process Development for Improved SRC Options: Interim Short Residence Time Studies," EPRI Report AF1158, Interim report, August 1979.

6) E. Gorin, C. J. Kulik, and H. E. Lebowitz, "Deashing of Coal Liquefaction Products Via Partial Deasphalting. 2. Hydrogenation and Hydroextraction Effluents," INEC Process Design and Developments, Vol. 16, Jan. 1977.

RECEIVED April 30, 1980.

Kinetics of Direct Liquefaction of Coal in the Presence of Molybdenum–Iron Catalyst

MINORU MORITA, SHIMIO SATOH, and TAKAO HASHIMOTO

Department of Chemical Engineering, Yamagata University, Yonezawa, Yamagata-ken 992, Japan

Many studies on direct liquefaction of coal have been carried out since the 1910's, and the effects of kinds of coal, pasting oil and catalyst, moisture, ash, temperature, hydrogen pressure, stirring and heating-up rate of paste on coal conversion, asphaltene and oil yields have been also investigated by many workers. However, few kinetic studies on their effects to reaction rate have been reported.

In this paper the effects of kinds of coal, pasting oil, catalyst and reaction temperature on coal liquefaction are illustrated, and a few kinetic models for catalytic liquefaction of five coals carried out in an autoclave reactor are proposed.

I. EXPERIMENTAL

Five coal materials were used in this study. These were Miike, Taiheiyo, Hikishima (Japanese coals), Morwell (Australian) and Bukit Asam (Indonesian) coals. The proximate and ultimate analyses of these coals are shown in Table 1. All of catalysts were powdered before

Table 1   Analyses of sample coals

Coal	Proximate analysis (wt%)				Ultimate analysis (wt%)		
	Moisture	Fixed carbon	Volatile matter	Ash	C	H	C/H
Miike	0.9	45.9	39.8	13.4	82.9	6.2	0.897
Hikishima	1.2	50.7	26.0	22.1	86.2	6.2	0.863
Taiheiyo	6.4	32.3	46.3	15.0	79.8	5.7	0.857
Morwell	12.6	52.4	34.2	0.8	65.3	5.2	0.956
Bukit Asam	9.5	44.8	45.0	0.7	68.6	5.2	0.910

use. In all of the experiments 30 wt% of powdered coal (passed through 100-mesh sieve) to paste, 0.033 of catalyst (weight ratio to the coal charged), a steel ball

0–8412–0587–6/80/47–139–213$05.00/0

(10 mm$^\phi$) and vehicle were charged to the 0.3 (for Miike
liquefaction) or 0.5 (for the other coal liquefactions)-
liter autoclave reactor in the required ratio. The
reactors were flushed and filled with cold hydrogen to
an initial pressure of 100 kg/cm^2-gauge at room temperature, heated to reaction temperature at a heat-up rate
of about 4°C per minite, held at constant temperature
for the desired length of time, and cooled to room
temperature at a heat-down rate of about 3°C per minite.
Then autoclave residues were treated with benzene and
n-hexane in Soxhlet apparatus, and the proportion of
"asphaltene" (defined as the benzene soluble and n-
hexane insoluble material), "oil" (the benzene and n-
hexane soluble material) in the liquefied product and
and "organic benzene insoluble" (the benzene insoluble
and organic material) was determined. Coal conversion is
defined as (1-organic benzene insoluble)x100/MAFcoal,
where MAF means moisture- and ash-free. Liquefaction
percent is defined as (asphaltene+oil)x100/MAFcoal.

II. RESULTS AND DISCUSSION

A. Effect of pasting oil on coal liquefaction

    1. Charged ratio of coal to pasting oil. Coal con-
version per cent on a moisture- and ash-free was independent of the charged ratio and had constant value
about 80-90 %, while liquefaction percentage decreased
with increasing the charged ratio. This result was
considered to be responsible for gasification with
thermal decomposition and coking of coals on inner wall
of the reactor; temperature at the wall would be higher
due to more viscous coal pastes in higher coal concentration. Therefore, well mixing was necessary to obtain a
good liquefaction percentage under higher coal concentration.
    2. Kind of pasting oil. Using four pasting oils
with boiling temperature of 330°C to 380°C, a coal
liquefied under the same reaction condition except for
the pasting oils. When a hydrogenated pasting oil was
used, a reaction rate was greater than with non-hydro-
genated pasting oils. This higher liquefaction rate for
the hydrogenated pasting oil was interpreted by the
action of greater proton-doner ability with it.

B. Effect of catalyst on coal liquefaction

    Figure 1 is the experimental result of Miike coal
liquefaction for MoO_3, $Fe(OH)_3$-S, $Fe(OH)_3$-MoO_3-S,

$H_2MoO_4 \cdot H_2O$ and $Fe(OH)_3$-$H_2MoO_4H_2O$-S catalysts. Figures 2 to 5 are those of Taiheiyo, Hikishima, Morwell and Bukit Asam coals for $Fe(OH)_3$-MoO_3-S. Figure 1-C shows reaction course for Miike coal under several reaction temperatures with nominal reaction time for $Fe(OH)_3$-MoO_3-S which is the most active among the catalysts. These results show that under temperature range 350°C to 410 °C a reaction rate increases with increasing temperature and a oil yield becomes greater with increasing nominal reaction time, whereas at the highest temperature 450°C a oil yield decreases, and both organic benzene insoluble (unreacted coal + coke) and asphaltene increase with increasing nominal reaction time. Reaction courses of Hikishima, Morwell and Bukit Asam coals for $Fe(OH)_3$-MoO_3-S (Fe-Mo-S) catalyst are shown in Figures 3 to 5, respectively. In Figures 4 and 5, it is shown that reaction rates and oil yields of Morwell and Bukit Asam coals are greater than any other tested coals at lower temperatures. Reaction courses for Taiheiyo and Hikishima coals, when the Fe-Mo-S catalyst was used, are shown in Figures 2 and 3. They show that at the highest temperature formed oil degrades to organic benzene insolubles in a similar way to that for Miike coal at the highest temperature. This characteristic was explained from forming of organic benzene insolubles by coking of produced asphaltene and oil. Degree of coking was dependent on both reaction temperature and kinds of catalyst, being considerable at the highest temperature. No coking was observed for $Fe(OH)_3$-S. Coking occures in the case of hydrogen lacking or unsufficient hydrogen diffusion conditions. Charged hydrogen weight ratio to coal paste was about 0.04 for Miike coal liquefaction and about 0.11 for the other coals, since Weller et al. ([1]) used about 0.08 of the ratio and Ishii et al. ([2]) used about 0.15, the ratio of 0.1 was sufficient to liquefy the coals. When the ratio was 0.04 for Miike coal, liquefaction however, might be in hydrogen poor state at reaction temperature 450°C. Greater trend of coking on Miike coal than the other coals might be partly responsible for this less hydrogen ratio. The reactant was agitated by the steel ball, and collision sound of the ball with a reactor wall could be heard except for runs in which the cokes was made remarkably. From both the fact of no agitating effect on a liquefaction rate, as shown by Maekawa et al.([3]), and such this easy moving of the steel ball, the agitation is adequate for the coal liquefaction, being not responsible for coking.

In Figures 6, 7 and 8 percentages of organic benzene insolubles to MAF coal charged in the reactors are plotted as a function of nominal reaction time on a

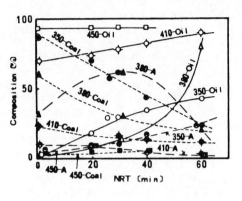

1A. Catalyst:MoO₃

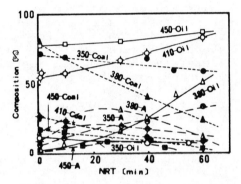

1B. Catalyst:Fe(OH)₃–S (weight ratio 1:1)

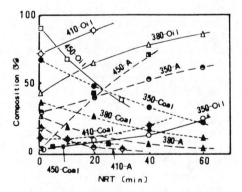

1C. Catalyst:Fe(OH)₃–MoO₃–S (1:1:1)

Figure 1.   Liquefaction courses of Miike

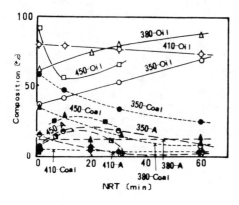

1D. Catalyst:$H_2MoO_4 \cdot H_2O$

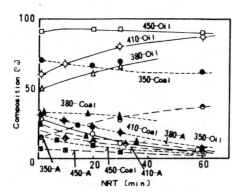

1E. Catalyst:$Fe(OH)_3 - H_2MoO_4 \cdot H_2O - S$
(1:1:1)

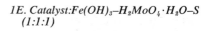

Temp.	Coal	Asph.	Oil
(°C)	(------)	(- - - -)	(——)
350	●	◐	○
380	▲	◭	△
410	⬗	⬖	◇
450	■	◪	□

NRT: Nominal reaction time

coal under various reaction temperatures

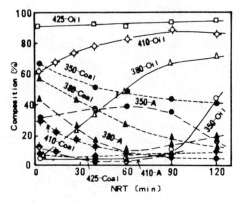

Figure 2.   Liquefaction courses of Tai-
heiyo coal

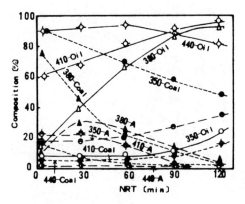

Figure 3.   Liquefaction courses of Hiki-
shima coal

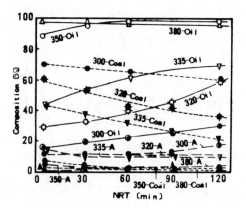

Figure 4.   Liquefaction courses of Mor-
well coal

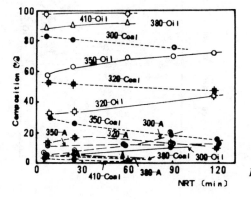

Figure 5.   *Liquefaction courses of Bukit Asam coal*

Temperature (°C)	Coal (-------)	Asphaltene (— — — — —)	Oil (————)
300	◉	◉	◉
320	✚	✜	⬓
335	▼	▼	▽
350	●	◓	○
380	▲	▲	△
410	✦	✦	◇
425	◆	◆	◇
440	◇	✦	◇

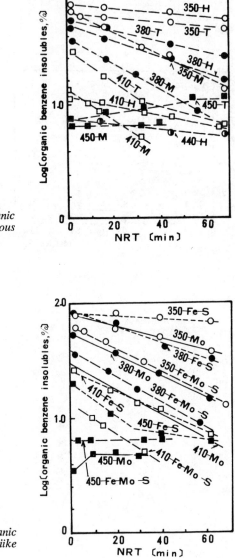

Figure 6. Log of percentages of organic benzene insolubles vs. NRT for various coals

Figure 7. Log of percentages of organic benzene insolubles vs. NRT for Miike coal

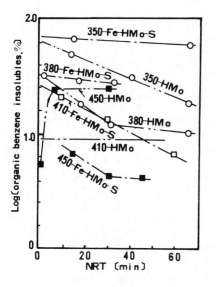

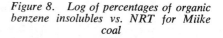

Figure 8. *Log of percentages of organic benzene insolubles vs. NRT for Miike coal*

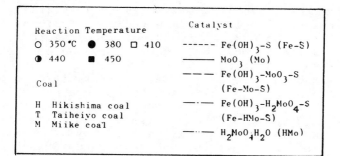

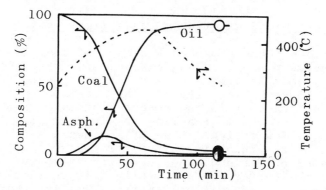

Figure 9. *Comparison of experimental liquefaction course for Miike coal with simulated one: (○, ●, ◑, – – –), experimental data; (———), simulated course. NRT = 16 min; reaction temperature = 450°C; catalyst = MoO₃.*

semilogarithmic graph paper in the same way as Ishii et
al. (2) showed in Sumiyoshi coal liquefaction. Reaction
rate is first order with respect to coal concentration,
since the plots at lower temperatures give straight
lines. At higher temperatures the plots do not give only
one decreasing straight line. This characteristic is
explained from greater extent of coking under these
temperatures. The specific reaction rates calculated
from the slope of lines are shown in Table 2.

Table 2 Rate constants (min^{-1}) on various catalysts

Catalyst	Reaction temperature (℃)			
	350	380	410	450
MoO_3	0.0145	0.0212	0.0253	–
$Fe(OH)_3$-S	0.0039	0.0113	0.0207	0.0338
$Fe(OH)_3$-MoO_3-S	0.0192	0.0253	0.0305	–
$H_2MoO_4 \cdot H_2O$	0.0188	0.0322	–	–
$Fe(OH)_3$-$H_2MoO_4 \cdot H_2O$-S	0.0023	0.0069	0.0230	–

(Coal:Miike)

From these results activities of the catalysts used
were compared. Conclusions are shown as follows:
  (a) MoO_3 is more active under lower temperatures,
      while $Fe(OH)_3$-S is more active under higher
      temperatures.
  (b) MoO_3 and $Fe(OH)_3$-S are complements each of the
      other, $Fe(OH)_3$-MoO_3-S being more active under
      both the lower and the higher temperatures.
  (c) $H_2MoO_4 \cdot H_2O$ of the catalysts containing water of
      crystallization have a tendency of resinifing
      and gasification, and $Fe(OH)_3$-$H_2MoO_4H_2O$-S is
      not so active as expected.
  (d) The action of the $Fe(OH)_3$-MoO_3-S catalyst to
      Taiheiyo coal is the same that to Miike coal.

C. Mechanism and kinetics of coal liquefaction

Various mechanisms and kinetics of coal liquefaction have been proposed and examined by many investiga
tors(1,2,4-8). As a general kinetic model of coal lique-
action, scheme 1 was assumed. The reaction rate of
every reaction step in the scheme assumed to be first
order with respect to reacting species and dissolved
hydrogen. A few typical cases of a general kinetic model
and the general characteristics for their cases are
illustrated on Table 3. When compared these typical
figures, the curves are apparently different in shape.
It is necessary to get the reaction data at lower
temperatures (i.e. 350-400°C) to distinguish between

Scheme 1 Kinetic Model of Direct Liquefaction of Coal

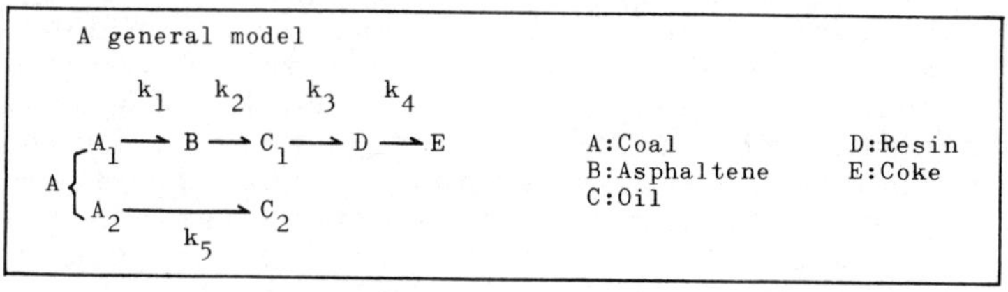

A general model

$$A \begin{cases} A_1 \xrightarrow{k_1} B \xrightarrow{k_2} C_1 \xrightarrow{k_3} D \xrightarrow{k_4} E \\ A_2 \xrightarrow{k_5} C_2 \end{cases}$$

A:Coal   D:Resin
B:Asphaltene   E:Coke
C:Oil

Table 3 Typical Cases of a General Kinetic Model

Case	Condition	Model	Coal	Catalyst	Worker
1	$A_1 \gg A_2 \doteqdot 0$ $k_1 > k_2 \gg k_3, k_4$	$A \rightarrow B \rightarrow C$	Pittzburg Seam	$SnCl_2-NH_4Cl$	S. Weller et al.
2	$A_1 \gg A_2 \doteqdot 0$ $k_2 > k_1 > k_4 > k_3$	$A \rightarrow B \rightarrow C \rightarrow D \rightarrow E$	Taiheiyo, Miike	$Fe(OH)_3-S,$ $H_2MoO_4 \cdot H_2O$	M. Morita et al.
3	$A_1 \gg A_2 \doteqdot 0$ $k_1 > k_2 > k_4 > k_3$	$A \rightarrow B \rightarrow C \rightarrow D \rightarrow E$	Miike	$Fe(OH)_3-$ MoO_3-S	M. Morita et al.
4	$A_1 > A_2$ $k_5 > k_1 > k_2$	$A_1 \rightarrow B \rightarrow C_1$ $A_2 \rightarrow C_2$	Yubari, Soya, Sumiyoshi	Red Mud	G. Takeya et al.
5	$A_1 > A_2$ $k_5 > k_2 \gg k_1$	$A_1 \rightarrow (B) \rightarrow C_1$ $(A_2) \rightarrow C_2$	Morwell, Bukit Asam	$Fe(OH)_3-$ MoO_3-S	M. Morita et al.

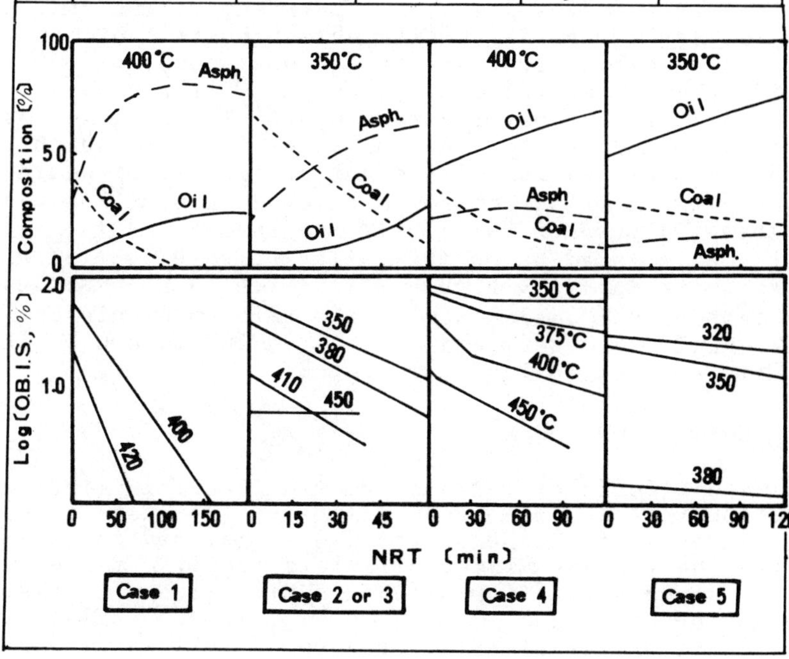

series reactions (Case 1, 2, 3) and parallel reactions (Case 4, 5), since for parallel reaction the time-logarithmic of organic benzene insolubles % curve does not give a straight line but two ones. A typical curve in Table 3 shows that when the apparent coal and asphaltene concentrations begin to increase gradually, a further series reaction, oil(C) $\longrightarrow$ resin(D) $\longrightarrow$ coke(E), should be assumed, where resin is defined as the materials soluble in benzene and insoluble in n-hexane and analyzed with asphaltene. The magnitude of the rate constant of each step is dependent on kinds of catalyst, and it is possible to find the catalyst which is very effective for promoting the reaction rates of any step in a kinetic model.

In this study the oil yield decreased with reaction time, as oil was polymerized at higher temperature for Miike, Taiheiyo and Hikishima coals. Thus the kinetic models (Case 2 or 3) which involve two steps of resinification and coking correlated data reasonably well for above coals, whereas for Morwell and Bukit Asam coals, Case 5 is more suitable.

Though conventional kinetic experiments are generally carried out with a autoclave at a high temperature and pressure, reaction is not isothermal but non-isothermal from the start of experiment to the end. As long as nominal reaction time which consists of the heat-up, constant temperature and heat-down periods is used, it will be difficult to estimate true rate constants. Therefore, the rate constants on Miike coal were estimated by the non-linear least square which involves minimization of the sum of squares of deviations between measured and calculated values. Validity of these values can be illustrated by agreement of course calculated using them with the experimental ones. To show the validity Figure 9 is given as an example. The temperature dependency of the rate constants on Miike coal was determined between 350°C and 450°C. The result is shown in Table 4.

III. CONCLUSIONS

The effects of various reaction conditions on the reaction rate and the mechanism of coal liquefaction were investigated. Conclusions are summarized as follows:
  (1) The reaction rate, oil yield are affected by kinds of pasting oil and ratio of coal to pasting oil.
  (2) Activities of catalysts are as follows:
      $H_2MoO_4 \cdot H_2O > Fe(OH)_3-MoO_3-S > MoO_3 > Fe(OH)_3-S$.
      The activity of $Fe(OH)_3-MoO_3-S$ may be due to the

Table 4 Rate constants calculated by non-linear least square method under non-isothermal condition

Catalyst	R.C. (min^{-1})	Reaction temperature ($°C$)			
		350	380	410	450
MoO_3	k_1	0.011	0.020	0.035	0.069
	k_2	0.050	0.092	0.158	0.306
	k_3	–	–	–	–
	k_4	–	–	–	–
$Fe(OH)_3-S$	k_1	0.008	0.015	0.026	0.051
	k_2	0.019	0.035	0.061	0.121
	k_3	–	–	–	0.003
	k_4	–	–	–	0.029
$Fe(OH)_3-MoO_3$ $-S$	k_1	0.018	0.029	0.047	0.082
	k_2	0.006	0.011	0.019	0.036
	k_3	0.–	–	0.001	0.007
	k_4	–	–	0.002	0.016
$H_2MoO_4 \cdot H_2O$	k_1	0.020	0.034	0.055	0.096
	k_2	0.050	0.110	0.223	0.519
	k_3	–	–	0.003	0.017
	k_4	–	–	0.010	0.057
$Fe(OH)_3-S-$ $H_2MoO_4 \cdot H_2O$	k_1	0.016	0.020	0.024	0.029
	k_2	0.081	0.128	0.193	0.318
	k_3	–	–	–	–
	k_4	–	–	–	0.003

(Coal:Miike)

concerted action with MoO_3 and $Fe(OH)_3$-S, and degrees of coking are dependent on kinds of catalyst.

(3) Under the same reaction conditions, the reaction rate are depend on the mechanism of coal liquefaction and kinds coal and catalyst. The reaction rate is in the following order: Morwell > Bukit Asam > Miike > Taiheiyo ≑ Hikishima coal.
The kinetic scheme of coal liquefaction would be expressed as follows:
for Morwell and Bukit Asam coals,

$$Coal \; {1 \atop 2} \underset{k_5}{\overset{k_1}{\rightleftharpoons}} Asphaltene \xrightarrow{k_2} Oil_1 \qquad\qquad Coal_1 > Coal_2$$
$$\searrow Oil_2 \qquad\qquad\qquad\qquad k_5 > k_2 \gg k_1$$

and for Miike, Taiheiyo and Hikishima coals,

$$Coal \xrightarrow{k_1} Asphaltene \xrightarrow{k_2} Oil \xrightarrow{k_3} Reain \xrightarrow{k_4} Coke$$

$$k_2 > k_1 > k_4 > k_3 \quad or \quad k_1 > k_2 > k_4 > k_3$$

The magnitude of each rate constant is depend on kinds of catalyst.

(4) In the kinetic analysis of the experimental data with an autoclave, the non-linear least square method was used to estimate the rate constants under nonisothermal conditions. The simulation of liquefaction calculated by substituing the estimated values into the rate equations showed good agreement with experimental values.

IV. REFERENCES

1. Weller, S., Pelipetz, M. G. and Frieman, S., *Ind. Eng. Chem.*, 43, 1572 (1951)
2. Ishii, T., Maekawa, Y. and Takeya, G., *Kagakukogaku (Chem. Eng. Japan)*, 29, 988 (1965)
3. Maekawa, Y., Shimokawa, K., Ishii, T. and Takeya, G., *Kogyokagaku-zasshi*, 73, 2347 (1970)
4. Falkum, E. and Glenn, R. A., *Fuel*, 31, 133 (1952)
5. Liebenberg, B. J. and Potgieter, H. G. J., *Fuel*, 52, (1973)
6. Guin, J. A., Tarrer, A. R., Taylor, Z. L. and Green, S. C., *Am. Chem. Soc. Div. Fuel Chem., Prepr.*, 20(1), 66 (1975), 21(5), 170 (1976)
7. Reuther, J. A., *Ind. Eng. Chem. Process Des. Dev.*, 16, 249 (1977)
8. Cronauer, D., Shah, Y. T. and Ruberto, R. G., *Ind. Eng. Chem. Process Des. Dev.*, 17(3), 281, 288 (1978)

RECEIVED June 4, 1980.

High-Yield Coal Conversion in a Zinc Chloride–Methanol Melt Under Moderate Conditions

JOHN H. SHINN[1] and THEODORE VERMEULEN

Chemical Engineering Department and Lawrence Berkeley Laboratory,
University of California, Berkeley, CA 94720

Converting coal to soluble material requires cleavage of enough chemical bonds to split the coal into subunits of only moderately high molecular weight. Because coal is inaccessible to conventional solid catalysts, current processing schemes use severe thermal conditions to effect the needed bond scission, forming some light gas and condensing some fragments into coke.

Hydrogen-donor action involving direct or indirect hydrogenation by solid catalysts has provided partial reductions in the severity of treatment, not sufficient to eliminate waste of coal and of input hydrogen. To lower the temperature sufficiently requires mobile catalysts which can penetrate the coal. Melts such as zinc chloride are therefore a promising medium.

Major work on zinc chloride catalysts for hydrogenation and hydrocracking of coal has been carried out by Zielke, Gorin, Struck and coworers at Consolidation Coal (now Conoco Coal Development Co.) (1). The emphasis there has been on a full boiling-point range of liquid product, from treatment at temperatures between 385 and 425°C and hydrogen pressures of 140 to 200 bars.

In this Laboratory, several potential liquid-phase treating agents have been studied at 225-275°C--that is, at temperatures well below 325°C, which appears to be the initiation temperature for pyrolysis of the coals studied here. Working with Wyodak coal in a $ZnCl_2$-water melt at 250°C, Holten and coworkers (2,3) discovered that addition of tetralin increased the pyridine solubility of product to 75%, compared to 25% without tetralin. About 10 wt-% of water is required in the melt, because pure $ZnCl_2$ melts at 317°C.

We have now found that replacing water in the melt by methanol leads to large increases in pyridine solubility of product from the treatment, even without tetralin addition. In this paper we characterize the effects of temperature, time, hydrogen pressure, reaction stoichiometry, and addition of various inorganic and organic additives. Because oxygen removal

[1] Present Address: Chevron Research Co., Box 1627, Richmond, CA 94804.

0–8412–0587–6/80/47–139–227$05.00/0
© 1980 American Chemical Society

from the coal occurs in parallel with solubilization, we conclude
that scission of ether-type C-O bonds is the primary chemical
reaction in either solubilization or liquefaction of subbituminous coal.

Experimental Procedures

The experiments were performed in a 600-ml Hastelloy B
stirred Parr autoclave fitted with a 300-ml glass liner.   275 gm
of $ZnCl_2$ (97+% pure from Matheson, Coleman, and Bell) was loaded
into the liner with the selected amount of methanol (Mallinkrodt
reagent-grade) and heated to about 150°C. At this time, 50 gm of
undried Roland seam Wyodak coal (-28 + 100 mesh) and, for some
runs, cosolvents were added to the melt. The autoclave was
closed, purged with hydrogen, and pressurized, so that it would
reach the desired hydrogen pressure at reaction temperature. The
contents were heated approximately 10°C/min with stirring until
the run temperature was reached. After reaction for a specified
period, the autoclave was immersed in a cold-water bath, de-
pressurized, and opened, and the contents were dumped into 2 1. of
cold water. The coal was then washed in a Buchner funnel with
6 1. of distilled water at 90°C, and dried to constant weight in
a vacuum oven at 110°C under 50 millibars of nitrogen. Some runs
were split after water quenching; in these, half of the product
was washed with dilute HCl before hot-water rinsing.
    A weighed portion of dried product (referred to as melt-
treated coal, MTC) was extracted to exhaustion sequentially with
benzene and pyridine in an atmospheric Soxhlet apparatus. The
extracts and residue were dried and weighed to determine the
solubility of the MTC.
    In addition to knowing the total MTC solubility, it was
important to determine the amount of methanol or other solvent
retained by the MTC. This quantity, the incorporation ratio (R,
gm incorporated organic material/gm coal-derived organic
material), was determined by a carbon balance on the reaction. By
assuming that any solvent retained in the dried MTC is pyridine-
soluble, and subtracting it from the total dissolved material, the
minimum solubility of the coal-derived material may be calculated.
This quantity, the corrected solubility, is an indicator of the
true solubilizing effect on the coal by the particular run conditions.
    Hydrogen consumption was measured by monitoring pressure
drops, and analyzing product gases in some runs. Additional
details on the experimental methods employed are available elsewhere (4).

Results

Effect of Reaction Conditions on Solubility. Earlier results (3) suggested investigation of the $ZnCl_2$-methanol system as
a coal-liquefaction medium based on high product solubility, low

incorporation, and relatively low cost of methanol.

Of primary concern were the effects of temperature, pressure, time, and methanol amount on the solubilizing activity of the melt. Figure 1 presents the effect of hydrogen pressure and temperature on corrected solubility. At 60 min reaction time with 50 gm of methanol, the solubility is roughly linear with hydrogen partial pressure. Even at 225°C there is significant conversion, with 800 psig producing 40% MTC solubility compared to 12% for the raw coal. By 275°C, conversion is rapid with nearly total solubility in one hour at hydrogen pressures as low as 200 psig.

As shown in Figure 3, solubilization roughly conforms to first-order kinetics, where rate = k[unconverted coal]. Rate constants of 3×10^{-2} and 1×10^{-1} min^{-1} are found for 250° and 275°C respectively, with nearly total conversion in less than 30 minutes at the higher temperature. Although negligible reaction takes place with heatup to 250°C (so-called "zero" time), considerable reaction occurs in the few minutes of heatup between 250° and 275°C. During this period, solubility rises 20%, incorporation approaches its maximum extent, and the H/C ratio drops to 0.75.

Effect of Reaction Conditions on Incorporation. In addition to improving solubilizing activity, it is desirable to limit the amount of methanol retained by the MTC. Table 1 shows that there is no significant effect of temperature on methanol incorporation at 800 psig, but a rapid rise in incorporation above 250°C at 200 psig hydrogen. Also, there is no trend in incorporation with time at 275°C, but a strong increase of incorporation with time at 250°C. As seen in Figure 4, there is less incorporation with 25 gm methanol than with 50 gm, and a leveling off of incorporation at higher hydrogen pressures.

Hydrogen-to-Carbon Ratios. An indicator of the quality of the MTC is the hydrogen-to-carbon ratio (raw coal H/C = 0.98). Figure 5 shows the effects of methanol amount and hydrogen pressure on the H/C ratio of the coal-derived portion of the MTC (Actual H/C ratios were corrected by subtracting the hydrogen in incorporated-CH_3 groups.). Higher hydrogen pressures result in higher H/C ratios regardless of methanol amount; higher methanol amounts produce lower H/C ratios. As seen in Figure 6, there is an initial drop in H/C ratios during the early stages of reaction, followed by a slow rise. This initial drop is larger and the subsequent rise more rapid at 275°C than at 250°C.

Effect of Additives. Table 2 lists the results of runs in which inorganic additives were used in the $ZnCl_2$-MeOH melt. Addition of 5 mole % ZnO had little effect on solubility and produced a surprising rise in H/C ratio of the MTC. The addition of 1 gm Zn powder had little effect in solubility, but slightly

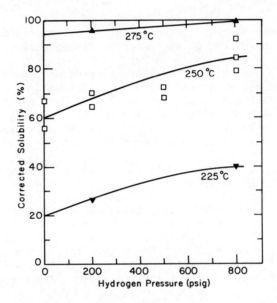

Figure 1.   Effect of hydrogen pressure and temperature on corrected total solubility: 273 g ZnCl₂, 50 g coal, 50 g CH₃OH; 60 min.

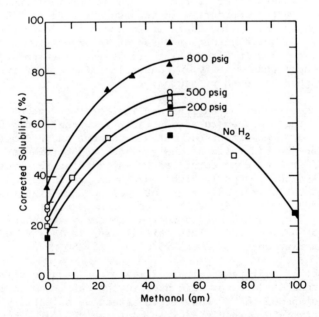

Figure 2.   Effect of methanol charge and hydrogen pressure on corrected total solubility: 273 g ZnCl₂, 50 g coal; 250°C; 60 min.

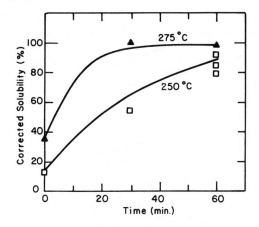

Figure 3.   Effect of run time on corrected total solubility at 250°C and 275°C: 273 g ZnCl₂, 50 g coal, 50 g CH₃OH; 800 psig H₂.

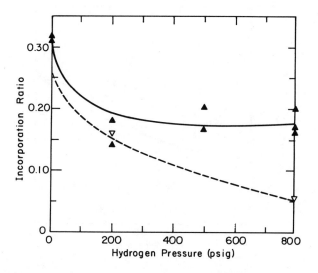

Figure 4.   Effect of hydrogen pressure and methanol charge on incorporation ratio: 273 g ZnCl₂, 50 g coal; 250°C; 60 min. Methanol: (▽), 25 g; (▲), 50 g.

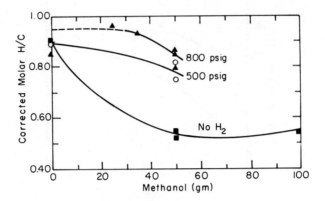

Figure 5.   Effect of hydrogen pressure and methanol charge on atomic H/C ratio: 273 g ZnCl₂, 50 g coal; 250°C; 60 min.

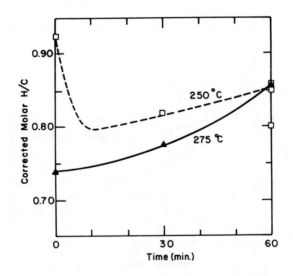

Figure 6.   Effect of run time on atomic H/C ratio at 250°C and 275°C: 273 g ZnCl₂, 50 g coal, 50 g CH₃OH; 800 psig H₂.

Table 1. Effect of Operating Variables on Incorporation of Methanol and on Corrected H/C Ratios (273 g $ZnCl_2$, 50 g Wyodak coal)

CH_3OH (gm)	H_2 (psig)	Temp (°C)	Time (min)	Corrected Solubility (%,daf)	Retained CH_3OH (gm/gm)[2]	Atomic H/C Ratio[3]	Retention per Unit Solubility
50	0	250	60	57.3[1]	0.31	0.53	0.60
	200			67.4[1]	0.16	0.81	0.20
	500			70.3[1]	0.19	0.78	0.25
	800			85.0[1]	0.18	0.85	0.22
25	200			54.8	0.16	0.70	0.28
	800			73.6	0.05	0.96	0.07
50	800	250	0	13.2	0.05	0.92	0.35
			30	53.7	0.16	0.82	0.28
		275	0	35.8	0.16	0.74	0.45
			30	100.0	0.21	0.78	0.21
	200	225	60	26.4	0.12	0.83	0.45
	800			40.0	0.11	0.88	0.27
	200	275		95.6	0.32	0.72	0.33
	800			99.1	0.17	0.86	0.17

[1] Average of replicate runs.
[2] Based on coal organics.
[3] Corrected for incorporation.

Table 2. Effect of Inorganic Additives to $ZnCl_2$-Methanol Melt. 273 gm $ZnCl_2$; 50 gm MeOH: 50 gm coal; T = 250°C; t = 60 min.

Additive (gm)	H_2 (psig)	Corrected Solubility (%, daf)	Retained CH_3OH[3] (gm/gm coal organics)	Corr. Atomic H/C
HCl (100 psig)	500	100.0	0.16	0.84
ZnO (9.0)[3]	500	69.2	0.11	1.10
Zn (1.0)	500	73.6	0.15	0.86
$CdCl_2$ (38.5)[1]	200	44.3	0.16	0.76
$SnCl_2$ (42.1)[2]	200	0.0	0.19	0.55
None[3]	500	70.3	0.18	0.79
None[3]	200	67.3	0.17	0.76

[1] 11.4 gm water present with $CdCl_2$.
[2] 36.0 gm water present with $SnCl_2$.
[3] Average of two replicate runs.

increased the H/C ratio. HCl, added to 100 psig pressure, while producing total solubility, had little more effect than a dilute HCl wash. The hydrated chlorides of tin and cadmium resulted in reduced yields.

Solvent additives to the melt (Table 3) fall into two categories: extractive and reactive. The extractive solvents (decane, perchloroethane, o-dichlorobenzene, and pyrrolidine) had negligible effect on solubility, possibly due to the preferential wetting of the coal by the solvent and exclusion of the $ZnCl_2$ melt. Reactive solvents (anthracene oil, indoline, cyclohexanol, and tetralin) all incorporated strongly. Donor solvents, tetralin and indoline, increase the "corrected" solubility, whereas anthracene oil and cyclohexanol have negligible effect.

Effect of Product Wash. For several runs, the product slurries were divided after water quenching of the MTC, and 15 ml HCl was added to the cold water wash. Figure 7 shows the increase in benzene and total MTC solubility as a result of the HCl wash. Acid-washing produces total pyridine solubility from a 65% soluble water-washed MTC. The effect of acid washing on benzene solubility is less marked, with a maximum increase of 10-15% when the water-washed benzene solubility is 25%. The maximum benzene solubility with either water or acid wash is 40%.

In some runs, a preliminary benzene wash was necessary to make the MTC sufficiently hydrophilic to allow removal of the $ZnCl_2$. The amount of solubilized material from the wash was added to the benzene Soxhlet yield, to give the total benzene solubility. As seen in Figure 8, relative to water-washed MTC, benzene-washed MTC has higher benzene solubility with the same total solubility, whereas HCl-washed MTC has higher total solubility with the same benzene solubility.

Oxygen Recovery and Solubility. Earlier work with the $ZnCl_2$ H_2O-system (3) had revealed a correlation between the oxygen recovery and the solubility of the MTC. There proved to be a similar relationship in the $ZnCl_2$-methanol system, as well as a separate relationship for acid-washed runs (Figure 9). Conditions of temperature, hydrogen pressure, reaction time and stoichiometry did not affect the relationship for a particular solvent, whereas additional solvent produced skew points supporting the conclusion that the relationship is solvent dependent. Acid washing produces increased solubility without affecting oxygen removal.

Discussion

Figure 10 shows the sequence of chemical and physical effects leading to coal solubilization under mild conditions. The first step in the conversion involves catalyst-coal contacting; a slow rate of contacting may limit the effective reactivity of the catalytic medium. The nature of the

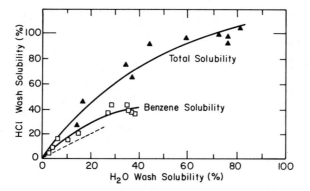

Figure 7. Effect of acid wash on solubilities compared with water wash: 273 g ZnCl₂, 50 g coal. Total solubility is the sum of benzene solubility and incremental pyridine solubility.

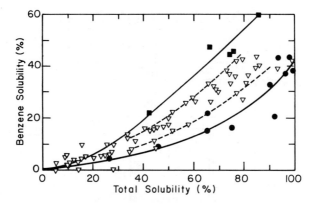

Figure 8. Relations between benzene solubility and total (pyridine) solubility: 273 g ZnCl₂, 50 g coal. Wash: (■), benzene; (▽), H₂O; (●), HCl.

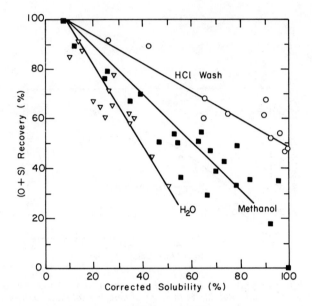

Figure 9.   Recovery of oxygen plus sulfur in product, relative to total solubility: 273 g ZnCl₂, 50 g coal.

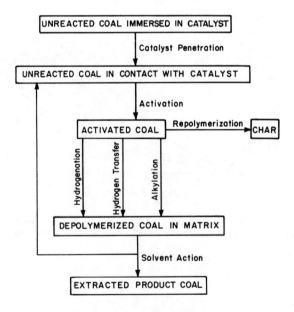

Figure 10.   Sequence of reactions leading to solubilization of coal under mild reaction conditions

Table 3. Effect of Solvent Additives in the $ZnCl_2$-Methanol Melt. 273 gm $ZnCl_2$; 50 gm MeOH; 50 gm coal; $T = 250°C$; t = 60 min.

Solvent (gm)	H_2 (psig)	Corrected Solubility (%, daf)	Total Retained Solvents (gm/ gm coal organics)
n-Decane (50)	250	41.4	0.19
C_2Cl_6 (50)	200	40.5	0.43
o-Cl_2-benzene (60)	200	33.5	0.16
Pyrrolidine (10.5)	500	65.6	0.13
Cyclohexanol (10)	200	68.7	0.39
Anthracene Oil (10)	250	73.7	0.77
Tetralin (10)	200	77.4	0.65
Indoline (10)	500	81.5	0.27
Methanol only	500	70.3	0.18

organic material in coal indicates that a reaction medium having a polar-organic character should be best able to penetrate the coal. Alcohols and phenols would appear desirable, particularly lower-molecular-weight alcohols which penetrate smaller pores. Methanol is likely to be important in aiding the $ZnCl_2$ penetration; extractive solvents may interfere by blocking the pores.

The second step involves coal activation. The relative ability of different media to split reactive crosslinkages of the coal is a crucial factor in obtaining conversion. The reactive crosslinks appear to be primarily ether bonds and aliphatic linkages, with suitably substituted neighboring aromatic centers (5,6). Work in these laboratories has shown that $ZnCl_2$ is an active catalyst for cleavage of these crosslinks (5,9). Addition of methanol may enhance this activity, whereas excessive solvent appears to dilute the catalyst.

Following activation, the cleaved weaker bonds must be properly "capped" to prevent polymerization to char. Several mechanisms are available for such capping. First hydroaromatic structures in the coal may exchange hydrogen with the reacted fragments, as noted by Whitehurst et al.(8). This type of donation may result in a net lowering of H/C ratio of the product as hydrogen is lost forming water upon oxygen removal. Two sources of external hydrogen are also available: donor solvents, and gas-phase hydrogen. The contribution of gas-phase hydrogen is normally small, but there is promise for enhancement of this effect through the use of additives with hydrogenation activity. Finally, capping may occur without hydrogen, by alkylation, or by alkoxylation with subsequent oxygen removal. Methanol appears important in this step, as its presence may prevent crosslinking subsequent to $ZnCl_2$ attack.

As solubilization may be impeded by inability of the catalyst to reach the reactive sites in the coal structure, a final process of solvation and removal of products by the reaction medium may play an important role. A medium which enhances physical disruption of the coal may enhance reactivity by increasing the reactant surface area, promoting intra-particular mass transfer, or making the initial products mobile for hydrogen shuttling and donation. Scanning electron micro-graphs of coal show that methanol addition causes massive physical changes in the coal particles, presumably enhancing phase contact and removing product during reaction so as to expose unreacted coal (7).

The effect of HCl washing in improving pyridine solubility may be explained in two ways. First, acid-base pairs may be dissociated by this wash, yielding two pyridine-soluble fragments from a pyridine-insoluble pair. Second, bivalent zinc ion may serve as a bridge between two high-molecular-weight fragments in the coal; such a bridged structure would be cleaved by HCl washing, producing $ZnCl_2$ and two fragments. Whether acid-base dissociation, zinc-bridge cleavage, or some other mechanism is responsible for the observed increase in solubility awaits further experimentation.

Comparison with Current Processing Techniques. It is appropriate to compare $ZnCl_2$- methanol catalyzed coal conversion with conventional thermolytic processing. Figure 11 presents in schematic form the conversion of coal to various products under different processing conditions. Pyrolytic processing, at 400–450°C, causes cleavage of many bonds in rapid succession. Distillable products may be formed directly in this manner, but the rapid rate of bond cleavage generally does not allow suitable capping, and significant quantities of char and gas are produced from condensation and fragmentation of the activated coal. Overly active catalysts (e.g., concentrated H_2SO_4, or $AlCl_3$) tend to give similar results.

In efforts to limit the side reactions, conventional processes utilize diverse methods of activating hydrogen, along with lower reaction temperatures (375-450°C). This moderate activation initially produces a depolymerized coal product consisting of mostly preasphaltenes, with some asphaltenes and oils; the total product has reduced nitrogen, sulfur, and oxygen levels relative to the original coal. This product may be used as a clean-burning boiler fuel, or may be reactivated by catalysts or hydrogen-rich recycle solvents to yield distillable products. Some char and gas formation still accompanies the moderate activation and reactivation steps, although signifi-cantly less than is formed by intense activation.

Our new catalytic route utilizes a mobile catalyst to contact the coal and perform the activation at still lower

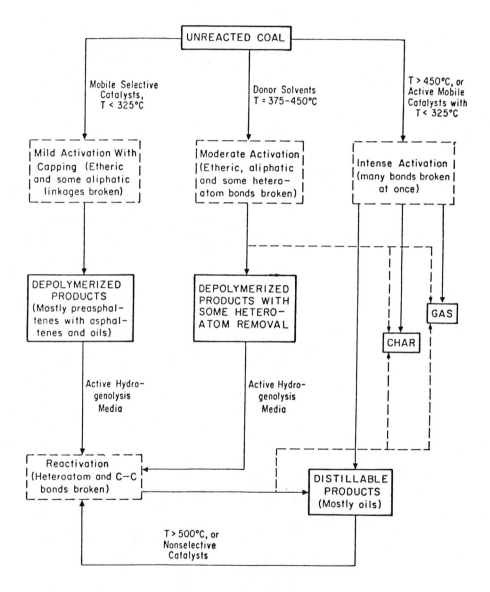

Figure 11. Sequential conversion of coal to distillable products, char, and gas under alternate reaction conditions

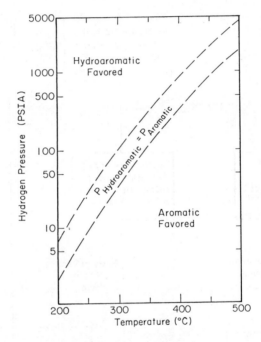

Figure 12.   Hydrogen pressure vs. reaction temperature for hydroaromatic–aromatic equilibrium.   Lower hydrogen pressures are needed for producing hydroaromatics in processing at lower temperatures.

temperatures, under still more controllable conditions. Suitable capping produces depolymerization products almost exclusively, avoiding the side reactions that form char and gas. Experiments in this laboratory suggest that this process may be extended (directly, or in a second stage), activating less reactive sites (C-O bonds, and others), by using the same $ZnCl_2$ catalyst at slightly higher temperature (e.g. 300°C), with hydrogen-donor solvents, higher hydrogen pressures, or hydrogenation co-catalysts.

A thermodynamic advantage allowing the use of lower hydrogen pressures accrues from use of lower temperatures for coal conversion. Figure 12 shows a plot of hydrogen pressure versus temperature for aromatic-hydroaromatic equilibria (benzene to cyclohexane, biphenyl to phenyl cyclohexane and bicyclohexyl, naphthalene to tetralin and decalin, phenanthrene to tetrahydro-, octahydro-, and perhydro-phenanthrene (10)). The region where the equilibrium concentration of aromatic and hydroaromatic are equal is shown as a diagonal band on the plot; the hydroaromatic form is thermodynamically favored above this region, while the aromatic is favored below. Thus, as the temperature is reduced, less hydrogen is needed to make dehydrogenation unfavorable. Since char formation results from dehydrogenation and condensation, a reduction in conversion temperature (which is accessible only with liquid catalysts) will allow lower hydrogen pressures to be utilized without threat of char formation.

Conclusions

We have discovered that $ZnCl_2$, in combination with methanol, constitutes an active liquid-phase catalytic medium for conversion of coal to pyridine-soluble material. There are several possible explanations for this effect: improved contact between coal and melt; higher activity of the $ZnCl_2$ in the methanol medium; methylation of cleaved bonds resulting in reduced char formation; and extraction of the reaction products leaving the coal more accessible.

Solubilization increases almost linearly with hydrogen pressure, at constant temperature and methanol charge. There is a strong effect of temperature, leading to complete solubility at 275°C in less than 30 minutes. Incorporation is best limited by using lower methanol ratios and higher hydrogen pressures.

Extractive solvents reduce the solubilization; donor solvents increase it, but involve incorporation. A relation between benzene and pyridine solubility is dependent on wash conditions. Finally, oxygen recovery and corrected solubility are related, the relationship varying with the solvent used.

Acknowledgment

The work reported was carried out under the auspices of the
Division of Chemical Sciences, Office of Basic Energy Sciences,
United States Department of Energy.

Literature Cited

1. Zielke, C. W.; Struck, R. T.; Evans, J. M.; Costanza, C. P.;
    Gorin, E. Ind. Eng. Chem. Proc. Des. Devel., 1966, 5, 158.

2. Holten, R. R.; Vermeulen, T. Univ. Calif. Lawrence
    Berkeley Laboratory, 1977, Report LBL-5948.

3. Shinn, J. H.; Hershkowitz, F.: Holten, R. R.; Vermeulen, T.;
    Grens, E. A., "Coal Liquefaction in Inorganic-Organic
    Liquid Mixtures". Paper presented at Am. Inst. Chem. Engrs.
    annual meeting, Miami, Florida, November 1977.

4. Shinn, J. H.; Vermeulen, T. Univ. Calif. Lawrence Berkeley
    Laboratory, 1978, Report LBL-9372.

5. Taylor, N. D.; Bell, A. T. Univ. Calif. Lawrence Berkeley
    Laboratory, 1978, Report LBL-7807.

6. Benjamin, B. M.; Raaben, V. F.; Maupin, P. H.; Brown, L. L.;
    Collins, C. J. Fuel, 1978, 57, 269.

7. Shinn, J. H.; Vermeulen, T. "Scanning Electron Microscopy
    Study of the Solubilization of Coal Under Mild Conditions".
    Paper submitted for presentation at Scanning Electron
    Microscopy Symposium, Washington, D.C., April 1979.

8. Whitehurst, D.D.; Farcasiu, M.; Mitchell, T. O.; Dickert,
    J. J., Mobil Res & Dev. Corp. "Nature and Origin of
    Asphaltenes in Processed Coals." Electric Power Research
    Institute, February 1976, Report AF-252.

9. Mobley, D. P.; Bell, A. T. Univ. Calif., Lawrence Berkeley
    Laboratory, 1978, Report LBL-8010.

10. Frye, C. G. J. Chem. Eng. Data, 1962, 7(4), 592.

RECEIVED May 21, 1980.

COMPOSITIONAL CHANGE ON COAL CONVERSION

The Effect of Coal Structure on the Dissolution of Brown Coal in Tetralin

R. J. HOOPER[1] and D. G. EVANS[2]

Department of Chemical Engineering, University of Melbourne,
Parkville 3052, Victoria, Australia

To describe in fundamental terms the dissolution of coal in a hydrogen-donor solvent requires an experimental approach that allows the chemical changes that occur within the coal during dissolution to be discussed. This, in turn, requires a direct method of determining the structural features in coal before it is reacted.

The solubilisation of coal by using the acid-catalysed reaction of coal with phenol is well recognised as an effective means of examining coal structure. The use of this technique to study the structure of an Australian brown coal from Morwell, Victoria, has previously been described by the authors (1). In that work, the coal was solubilised by reacting it with phenol and then separated, although other subfractions were also isolated, into four major fractions; the first rich in aliphatics, the second rich in simple aromatics, the third rich in di-aromatics and polar groups, and the fourth rich in poly-aromatics. These fractions can be regarded as models of structural types within the coal and by reacting each fraction separately the role played by different chemical structures during the hydrogenation process can be examined - allowing the direct study of the effect that chemical type has on the coal hydrogenation reaction.

In the work now reported coal fractions derived from a solubilised coal were reacted individually with Tetralin, without any additions of catalyst or gaseous hydrogen, and the reaction products studied to determine the effect that chemical type had on the reaction. The untreated whole coal was also reacted to test whether phenol, present in the coal fractions as a result of the fractionation procedure, was having any significant effect on the reaction with the fractions.

[1] Current address: Liquid Fuels Trust Board, Box 17, Wellington,NZ.
[2] Current address: Centre for Environmental Studies, University of Melbourne, Parkville 3052, Victoria, Australia.

Experimental

Experimental Procedure. Morwell brown coal was solubilised
by reacting with phenol, in the presence of para toluene sulfonic
acid, at 183ºC, and the reaction product was then separated into
four fractions and analysed according to procedures described
elsewhere (1). The structural characteristics of the four
fractions as determined by the present work and confirmed by
reference to the literature (2,3) are summarised in Table I. As
these characteristics are influenced to some extent by the
presence of chemically combined phenol, the content of this in
each fraction is also estimated.

Approximately 3g samples of the coal fractions and of the
whole coal were then reacted separately with 25 - 30 ml of tetralin at 450ºC in a type 316 stainless steel, sealed reactor, 13 cm
high by 2 cm diameter. The reactor was heated by plunging it
into a preheated fluidised sand bath; after 4 hours it was
removed and quenched rapidly.

The temperature of the reaction mix was measured by a
stainless steel-sheathed thermocouple inserted through the
reactor cap. Heating up and cooling down times were small
compared with the total reaction time. In all cases the free
space in the reactor was flushed with nitrogen before sealing,
and the reaction proceeded under a small initial nitrogen
pressure.

After reaction, any solid residue was filtered off and the
liquid product was separated by distillation into a bottoms
product and a distillate that included unreacted Tetralin and low-
boiling products from both the coal and the Tetralin. As
tetralin breaks down under dissolution conditions to form mainly
the tetralin isomer 1-methyl indan, naphthalene and alkyl
benzenes (4) it was assumed that no compound with a higher
boiling point than naphthalene was formed from the solvent, and
the distillation to recover solvent was therefore continued
until naphthalene stopped subliming. Some residual naphthalene
remained in the bottoms product; its mass, as determined from
nmr and elemental analysis, was subtracted from the mass of
bottoms product recovered and included in the amount of
distillate recovered. It was assumed that all naphthalene present
came from the Tetralin, not the coal. However, as the amount of
tetralin reacted was 10 times the amount of coal this
assumption appears reasonable.

Material formed from the coal which appears in the distillate
is here called solvent-range material, following the terminology
used by Whitehurst et al. (5). Its mass was estimated by mass
balance over the material recovered from the reaction as it could
not be separated from the large excess of Tetralin and Tetralin
breakdown products also contained in the distillate. This procedure includes with the mass of solvent-range material any gases
and water formed in the reaction.

TABLE I. **Structural Characteristics of Coal Fractions Separated from Solubilized Brown Coal (1)**

Fraction	Mass % Phenol	Structural Characteristics
A A liquid soluble in pentane	65	Mostly aliphatic material with some mono-aromatic parts broken off the coal by C-C cleavage. Apart from combined-phenol it has negligible polar material. It contains some free paraffinic material, but exists mostly as alkyl phenols and alkyl-aryl ethers
B A liquid insoluble in pentane but soluble in benzene	65	A mixture of alkyl side chains and aromatic fragments, predominantly di-aromatic. It exists either as alkyl phenols or as aromatic fragments attached to phenol by methylene bridges. It also contains other oxygen functional groups
C A pitch in-soluble in benzene but soluble in benzene/ethanol azeotrope	40	Consists almost entirely of aromatic fragments attached to phenol by methylene bridges. These fragments are larger than in fraction B as they contain polyaromatic groups. It has more oxygen functional groups than B.
D A solid insoluble in ethanol/benzene azeotrope	25	Predominantly diaromatic and poly-nuclear perhaps combined through naphthenic bridges, with negligible aliphatic content.

Analysis Techniques. The contents of the major breakdown products of tetralin (naphthalene and 1-methyl indan) present in the distillate were determined by gas-liquid chromatography using a Hewlett Packard Series 5750 Research Chromatograph with a 62m x 0.5mm diameter glass capillary SCOT column coated with non-polar SE 30 liquid phase (see Reference (4) for details).

Infrared spectra of the original coal, the original coal fractions, and all bottoms product and residues derived from them were measured on a Perkin Elmer 457 Grating Infrared Spectrophotometer. Liquid samples were analysed as a thin film or smear. Solid samples were prepared as a KBr disc containing approximately 0.3% by weight sample. The disc was prepared by grinding the KBr mixture for 2 minutes in a tungsten carbide TEMA grinding barrel, drying for 24 h in a vacuum desiccator over phosphorus pentoxide, then pressing into a disc at 10 tons force, at room temperature, but under vacuum.

Proton nmr spectra of fractions A, B and C and all bottoms products were recorded on a Varian HA 100nmr spectrometer using a solution of the sample dissolved in pyridine-d$_5$. Spectra were run at room temperature with tetra methyl silane (TMS) as an internal standard, with a sweep width of 0 to 1000 cps from TMS. Fraction D and the whole coal were only partly soluble in pyridine and it was therefore not possible to get representative spectra from them.

Carbon, hydrogen and oxygen contents of the original coal, original fractions, bottoms and residues were determined microanalytically by the CSIRO Microanalytical Service. Ash contents of samples were determined in a standard ashing oven (6). Phenolic and carboxylic oxygen contents were measured by the State Electricity Commission of Victoria using techniques developed by them for brown coals (7).

Results

Recovery of Coal Material from the Reaction with Tetralin. The yields of the different products from the reactions of the various fractions with tetralin are summarised in Table II.

These yields are also given on the basis of 100 g of original dry coal before fractionation. The bottom line of the table shows the mass of each fraction obtained from 100 g of dry coal. For every 100 g of original dry coal an additional 100 g of extraneous material was present. Elemental balances and other evidence (1) showed this to be made up almost entirely of phenol chemically combined with the coal material, with traces present of residual solvent associated with the fractions as a result of the coal preparation and fractionation scheme. Note that with fractions A and B no solid residue was obtained.

The amount of residue recovered from the other two fractions is almost the same as that recovered from the whole-coal reaction, suggesting that the combined phenol and residual solvent end up completely in the bottoms product and solvent-range product.

TABLE II

Yields of Original Coal Fractions and their Products of Reaction with Tetralin, g/100g Original Dry Coal.

Comp-onent	Fraction									Whole Coal
	A		B		C		D		Compo-site	
	%of frac-tion	g/100g dry coal	%of frac-tion	g/100g dry coal	%of frac-tion	g/100g dry coal	%of frac-tion	g/100g dry coal	g/100g dry coal	g/100g dry coal
bottoms	47	13	56	37	37	29	30	9	88	46
residue	0	0	0	0	27	21	61	17	38	40
solvent range	53	15	44	29	36	28	9	2	74	14
total fract-ion	100	28	100	66	100	78	100	28	200	100

Composition of the Coal Products. Table III shows elemental compositions of the original coal fractions, the solid residues and the bottoms products, together with the portions of the total oxygen present in the original fractions as phenolic and carboxylic groups. Because of dilution with combined phenol and residual solvent the composite analysis of the original fractions has higher carbon and lower oxygen contents than the original whole coal.

As expected, both the bottoms products and the residues, where formed, have substantially higher carbon and lower oxygen contents than the original fractions, but whereas in the bottoms products the hydrogen contents have increased, in the residues they are reduced. The bottoms products, including that from the whole coal, are remarkably similar in composition to each other. Lkkewise the residues are similar in composition to each other.

Figure 1 shows representative infrared spectra for the fractions before reaction and for bottoms products and residues. Although there were considerable differences in the spectra of the four original fractions the spectra obtained for their bottoms products were quite similar. The spectra show that significant amounts of aliphatic material (2850 and 2920 cm $^{-1}$) are present in the bottoms. An aromatic content is indicated by

**TABLE III.   Composition of the Fractions and Their Reaction Products, Mass %.
N.D. Means Not Determined (in the Case of the Residue from Fraction C Insufficient Sample Was Available for an Ash Determination).** *Note* **That the Method for
Backing Out Naphthalene from the Bottoms Involves Normalizing the Composition
to C + H + O = 100%.**

| | Fraction | | | | | Whole Coal |
	A	B	C	D	composite	
original fraction						
C	76	74	69	71	72	63
H	7	6	5	4	5	5
O phen- olic	4	4	3	6	4	5
O carb- olic	0	1	3	2	2	5
O total	17	16	21	18	18	25
ash	-	-	2	3	2	4
unacc- ounted	1	4	3	4	3	3
residue						
C			89	86	88	85
H			4	4	4	4
O			3	3	3	3
ash			N.D.	7	N.D.	8
unacc- ounted			N.D.	0	N.D.	-2
bottoms						
C	86	83	86	83	85	85
H	7	7	7	7	7	8
O	7	10	7	10	8	7

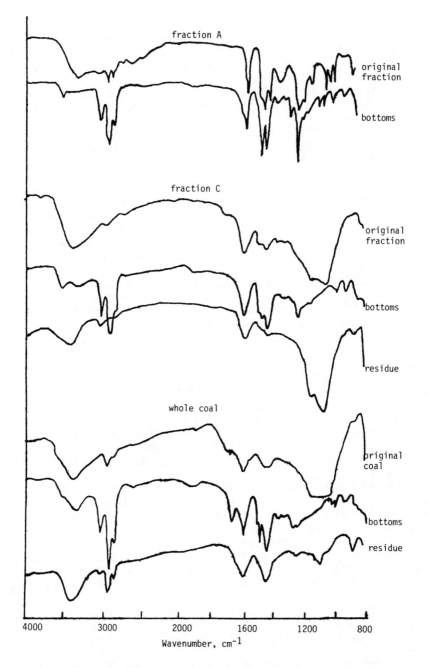

Figure 1.   IR spectra for typical fractions and their products

the aromatic C-H stretching vibration at 3030 cm-1, but this is
due in part, at least, to residual naphthalene and to phenol
combined with the original fraction. Absorption at 3400 cm-1
(due to hydrogen bonded OH), present both in the coal fractions
and the coal before reaction, has almost disappeared in the
products from the coal fractions. The absorption does still
occur in the bottoms product from the whole coal, although it is
greatly reduced.

The spectra of the bottoms products show that absorption due
to carbonyl groups (at 1700 cm-1) and the broad absorption in the
region 1000 - 1200 cm-1 due to oxygen functional groups, both
normally present in coal, are no longer present, indicating, as
would be expected, that functional groups are destroyed during
the reaction. The absence of a large hydrogen-bonded OH peak at
3400 cm-1 indicates that the remaining oxygen absorption at
1250 cm-1 is not due primarily to phenol. This absorption may
be due to ether, but this assignment is by no means certain, as
normally ether absorption in this region is broad whereas the
spectra show sharp absorptions.

Aliphatic material still remains in the residue from the
whole coal, but is virtually eliminated in the residue from
fraction C. The absorption at 1170 cm-1 in the spectra of both
residues may be due to benzofuran type structures (8), but it is
felt that the strong absorption in the region 1000 - 1200 cm-1
may have been enhanced by the presence of silica, a major
component of the ash content in this coal.

Table IV shows the proton ratios obtained from the nmr
spectra on the original fractions A, B and C and all the bottoms
products. The proton ratios for the bottoms products have been
adjusted to eliminate absorptions due to residual naphthalene.
Note that the original fractions contained hydrogen present from
combined phenol. Most of this hydrogen appears as monoaromatic
hydrogen, but when the phenol-OH is still intact, one proton will
appear as OH hydrogen.

The nmr analyses of the bottoms products given in Table IV
show the material to have a large aliphatic content. The
aromatic/aliphatic ratios of the fractions are higher than for
the whole coal because of the presence of combined phenol;
reaction with Tetralin reduces these ratios considerably,
presumably by transfer of much of this material to the solvent-
range product, but some of it must remain in the bottoms as the
aromatic/aliphatic ratio of the composite bottoms product from
the fractions is higher than that from the whole coal. It was
not possible to calculate the contribution that the diluents,
excess solvent and combined phenol, made to the aromatic H, but
the large monoaromatic content of the bottoms product must be due,
in part, to these.

The remarkable feature in Table IV is that after the spectra
have been adjusted for naphthalene, none of the bottoms products
show the diaromatic or polyaromatic material which were present

TABLE IV

Ratio of protons present in various forms to total protons, %	A		B		C		D	composite	whole coal
	orig bottoms fraction	bottoms	orig bottoms fraction	bottoms	orig bottoms fraction	bottoms	bottoms	bottoms	bottoms
OH hydrogen	7	0	17	0	19	0	0	0	0
polyaromatic	0	0	0	0	6	0	0	0	0
diaromatic	6	0	10	0	8	0	0	0	0
monoaromatic	63	46	58	54	44	42	43	48	33
olefinic	1	0	0	0	0	0	0	0	0
methylene bridge	4	5	9	4	17	1	1	3	0
α methine and methylene	3	15	0	17	0	21	21	18	24
α methyl	3	8	1	8	4	11	11	9	8
β methylene	13	10	1	13	0	14	16	13	20
β methyl	0	12	4	5	1	10	7	8	12
γ aliphatic	0	4	0	0	0	0	0	1	4
H_{ar}/H_{al}	2.8	0.9	4.5	1.2	2.7	0.7	0.8	0.9	0.5

Distribution of protons by type and overall aromatic/aliphatic proton ratios for the original fractions and bottoms products, as determined by proton nmr. Proton distribution for fraction D and the whole coal are not included as these materials were only partly soluble and the resultant spectra were not representative of the whole material.

in the original materials. It is possible that the presence of
naphthalene identified in the product may have masked any
absorptions from diaromatic species present in the material,
however, the nature of the spectra allowed the presence of
naphthalene to be readily accounted for and any possible contribution from diaromatics is considered small. In the case of
fractions C and D one might expect the diaromatic and polyaromatic
material to end up in the solid residue, but in the reactions of
fraction A and especially with fraction B, where a large
diaromatic content existed before reaction, no residue was
formed, thus suggesting that aromatic rings must be broken during
the reaction. In addition, none of the OH hydrogen present in
the original coal fractions appears in the bottoms product.

The solvent-range product was not separately analysed as it
was not able to be separated from the recovered solvent in the
distillate. However, GLC examination of the distillate indicated
that the solvent-range product was derived mainly from aliphatic
side chains in the coal (9). Note that virtually no solvent-
range product was derived from fraction D.

Naphthalene and 1-methyl indan contents in the distillate
were determined in order to calculate the amount of hydrogen
transferred to the coal material from the solvent. The hydrogen
transferred to each fraction has been calculated in Table V in
terms of 100 g of the original coal before fractionation and this
shows a composite value of hydrogen transferred to the whole coal
of 9.5 g/100g of original coal compared with 3.6 g/100g of
original coal when measured directly from a reaction of the whole
coal. When adjusted for the amount of hydrogen that would be
consumed by chemically attached phenol present in the fractionated material the calculated composite value reduces to 7.1 g/100g
of original coal; still well above the hydrogen transferred to
the unfractionated coal. It is thus apparent that the
fractionation procedure has enhanced the hydrogen transfer
process, presumably by providing additional sites for hydrogen
transfer to the coal. Further, when adjusted for the masses of
chemically combined phenol present in the individual fractions,
the weight of hydrogen transferred per 100 g of coal material
contained in each fraction (as shown in Table V) indicates that
hydrogen transfer to the coal occurs to the greatest extent to
that part of the coal represented by fraction C. This fraction
contains the greatest content of oxygen functional groups.

Discussion

The Effect of Phenol. Three types of phenol compounds have
been identified in the fractions derived from the product of the
phenolation reaction (1,2): alkyl phenols and alkyl-aryl ethers,
both formed by combining phenol with alkyl side chains cleaved
from the coal molecule, and compounds made up of aromatic
fragments attached to phenol by a methylene bridge, formed by

TABLE V
Calculation of hydrogen transferred to the coal fractions, g/100g of dry brown coal

| | Fraction | | | | | |
	A	B	C	D	Composite	Whole coal
Weight of fraction recovered, g/100g of dry brown coal	28	66	78	28	200	100
Weight of H transferred, g/100g of fraction	2.8	3.4	6.2	5.2	–	3.6
Weight of H transferred, g/100g of dry brown coal	0.8	2.3	4.9	1.5	9.5	3.6
Weight of H transferred, adjusted for phenol, g/100g of coal material	4.1	5.8	9.0	6.3	–	3.6
Weight of H transferred, adjusted for phenol, g/100g of dry brown coal	0.4	1.3	4.1	1.3	7.1	3.6

cleaving aromatic-aliphatic linkages in the coal and exchanging
the aromatic structures with phenol. For the hydrogenation of
coal fractions separated from the phenolated product to simulate
the hydrogenation of the whole coal the removal of coal fragments
from the coal molecule during the phenolation reaction must
involve similar C-C cleavage processes as would occur by the
thermal breakdown of coal during a hydrogenation reaction. For
example, one would expect alkyl side chains to be cleaved off
under hydrogenation conditions, in much the same way as has been
seen to occur in the phenolation reaction. Moreover, other
workers have shown that the molecular weights of coal fragments
from the phenolation reaction are in the region 300 - 1000 (3)
which is the same molecular weight range as for products from a
coal dissolution reaction (4). Thus, both processes, one
involving C-C cleavage by phenolation, the other by thermal
breakdown, produce coal fragments of the same size. The
hydrogenation of the coal fragments can therefore be considered to
simulate the reaction of the whole coal, providing appropriate
allowance is made for the movement of the phenol groups themselves.

   The Nature of the Products. The composition of the bottoms
products from the various reactions were all similar, regardless
of the original material. Elemental composition ranged from 83.3%
to 86.4% carbon, from 6.6% to 7.2% hydrogen, and from 6.7% to
10.1% oxygen. Infrared and nmr analysis showed the material to
contain a significant aliphatic content, with hydrogen bonding in
the product almost destroyed. The large monoaromatic hydrogen
content of the bottoms product (see Table IV) must be due in part
to hydrogenation products of combined-phenol remaining from the
dissolution reaction. The contribution of aromatic H due to the
diluents in the bottoms product was not able to be measured as
hydrogenation products from the combined-phenol will also end up
in the solvent-range material.

   Similarly, the residues which appear as a carbon-rich material
with very little oxygen, were all alike. Their elemental
compositions ranged from 86.1% to 89.3% carbon, 3.8% to 4.2%
hydrogen, and 2.6% to 2.9% oxygen. The material is mainly
aromatic with perhaps some benzofuran type structures, suggesting
that condensation reactions may be involved in its formation.
Most importantly, the solid residue was shown to form only from
fractions C and D. As fraction C was completely soluble in tetralin, and the reactor feed before reaction was therefore liquid,
the solid material present after reaction was, in the case of
fraction C at least, not present in the original coal and must
therefore be a product of the reaction.

   Although the usual nomenclature in calling this solid a
"residue" has been followed, such nomenclature is misleading in
terms of reaction mechanism. Some of the "residue" formed in the
reaction of the whole coal is genuine unreacted residue and some
is a reaction product with the evidence suggesting that
condensation reactions may be involved in its formation (10).

The Coal Dissolution Process. The classic view of the mechanism for the reaction between coal and a hydrogen-donor solvent involves the thermal breakdown of the single carbon-carbon bonds within the coal to produce reactive fragments in the form of free radicals which are then stabilised by hydrogen transferred from the solvent or elsewhere in the coal. If insufficient hydrogen is available the aromatic fragments can polymerise yielding chars or coke. Another view of coal dissolution less widely held, is that oil is produced from coal via reaction pathways involving asphaltene intermediates (11).

A most striking result from the work described above is that the composition of the bottoms product and residues from the dissolution reaction did not depend on the chemical structure of the original coal material; only their relative quantities differed. This supports the view of a mechanism involving the stabilisation of reactive fragments rather than an asphaltene-intermediate mechanism. The formation of a carbon-rich condensed material as a residue of the reaction and the fact that hydrogen transfer occurred largely to specific parts of the coal further supports this view.

Perhaps the greatest difference between the present work and more conventional work on higher-rank coals is the important role of the functional-group oxygen. The importance of oxygen groups is stressed, as the predominant process during the dissolution reaction was the destruction of functional groups within the coal. The role of oxygen in the reaction was not clearly defined by the present work, but quinone groups, for example, are thought to play a role in the liquefaction of high-oxygen-content coals (12), and their effectiveness in the free radical abstraction of hydrogen from hydrogen-donor compounds is well known (13).

Acknowledgements

We thank the New Zealand DSIR and the Australian Research Grants Committee for supporting this work.

Abstract

Four fractions representing chemical types present in the original coal were separated from a Victorian brown coal solubilised by reacting with phenol in the presence of para toluene sulphonic acid catalyst. These fractions, plus a whole coal sample, were each reacted separately with tetralin in a nitrogen atmosphere for 4h at 450°C. Three products were recovered from the reaction: a solvent-range product (material derived from the coal having the same boiling point range as the tetralin solvent) a liquid bottoms product with boiling point greater than the solvent, and a solid residue. The structural features of the bottoms product and the residue from the different reactions were determined from elemental and spectroscopic analysis. It was shown that the yield of the various products depended markedly on chemical structure within the coal. However, the composition of the bottoms product did not depend on the chemical structure of the original material. The dissolution of brown coal in Tetralin was best described by a free radical reaction involving the destruction of oxygen functional groups in the coal material by hydrogen transfer from the solvent. Condensation of aromatic coal liquids to a solid carbon-rich condensed material was also shown to occur under the reaction conditions.

Literature Cited

1. Evans, D.J., Hooper, R.J., Prepr., Fuel Chemistry Division, ACS/CSJ Congress, Honolulu, April 1979; Adv.Chem.Ser., To be published.
2. Heredy, L.A., et al., Fuel 1962, 41, p221
3. Heredy, L.A., et al., Fuel 1965, 44, p125
4. Hooper, R.J., Battaerd, H.A.J., Evans, D.G., Fuel 1979, 58 p132
5. Whitehurst, D.D., et al., EPRI AF-480, EPRI, California, July 1977.
6. B.S. 1076, Part 3, 1973
7. McPhail, I., Murray, J.B., Victoria, State Electricity Commission, Scientific Div. Rep. MR-155, 1969
8. Ouchi, K., Fuel 1967, 46, p319
9. Hooper, R.J., "The Dissolution of a Victorian Brown Coal in a Hydrogen-Donor Solvent", PhD Thesis, University of Melbourne, Australia, April 1978
10. Hooper, R.J., Evans, D.G., Fuel 1978, 57, p799
11. Yoshidi, R., et al., Fuel 1976, 55, p337,341
12. Brower, K.R., Fuel 1977, 56, p245
13. Braude, E.A., et al., Journal Chem Soc, 1954, p3574.

RECEIVED March 28, 1980.

14

Coal Liquefaction Under Atmospheric Pressure

ISAO MOCHIDA and K. TAKESHITA

Research Institute of Industrial Science, Kyushu University 86, Fukuoka 812, Japan

Liquefaction of coals has been extensively investigated in the recent time to synthesize liquid fuels
of petroleum substituent (1). The processes for liquefaction proposed are classified into three major
groups. They are direct hydrogenation of coal under
high hydrogen pressure, the solvent refining of coal,
and the hydrogenation of liquid produced by dry distillation of coal. Among them, the solvent refining may
be the most skillful method for the largest yield of
coal liquefaction under the moderate conditions (2).
This process may be further subdivided into two categories. First idea consists of extraction of coal
molecules, using suitable solvents such as anthracene
oil or toluene under super critical conditions (3).
Another one depends on the moderate hydrogenation of
coal molecule using solvents of hydrogen transfer
ability such as hydrogenated anthracene oil (4) and
tetraline (5) under medium hydrogen pressure.

The present authors studied the solvolytic liquefaction process (6,7) from chemical viewpoints on the
solvents and the coals in previous paper (5). The
basic idea of this process is that coals can be liquefied under atmospheric pressure when a suitable
solvent of high boiling point assures the ability of
coal extraction or solvolytic reactivity. The solvent
may be hopefully derived from the petroleum asphaltene
because of its effective utilization. Fig. 1 of a
previous paper (8) may indicate an essential nature of
this process. The liquefaction activity of a solvent
was revealed to depend not only on its dissolving
ability but also on its reactivity for the liquefying
reaction according to the nature of the coal. Fusible
coals were liquefied at high yield by the aid of
aromatic solvents. However, coals which are non-
fusible at liquefaction temperature are scarcely

0–8412–0587–6/80/47–139–259$05.00/0

liquefied with the non-reactive solvent. This fact
indicates the importance of solvolytic reactivity in
the liquefaction such coals. This conclusion corresponds to the fact that tetraline or hydrogenated anthracene oil assured the high liquefaction yield regardless of the coal ranks (4,5), although the processes
require some high pressure.
     In the present study, the liquefaction activities
of pyrene, its derivatives, and decacyclene with coals
of several ranks are studied to ascertain the previous
ideas of liquefaction mechanism and to develop novel
liquefaction process under atmospheric pressure. The
coals used in the present study are non-fusible or
fusible at relatively high temperature, and then gave
small liquefaction yield with pyrene of a non-solvolytic solvent at 370°C.

Experimental

     Coals. The coals used in the present study are
listed in Table 1, where some of their properties are
also summarized. They were gratefully supplied from
Nippon Steel Co., Nippon Kokan Co., and National
Industrial Research Laboratory of Kyushu.

     Liquefaction Solvents. The solvents used in the
present study are listed in Tables 2 and 3. Alkylated
and hydrogenated pyrenes were synthesized by Friedel-
Crafts and Birch reduction, respectively. Details have
been described in another place (9).

     Procedure and Analysis. Apparatus used in this
experiment consisted of a reactor of pyrex glass(diameter 30 mm, length 250 mm, volume 175 ml) with a stirring bar and a cold-trap. After 1~3 g of coal and
described amount of the solvent were added in the
reactor, of which weight was previously measered, the
reactor was heated in a vertical electric furnace under
N₂ gas flow. The temperature was increased at the rate
of 4°C/min, and was kept at the prescribed temperature
for 1 hr. The weight calculated by substracting the
weight of the reactor from the total weight was defined
to be residual yield (remaining coal and solvent). The
weight of oil and sublimed matter captured in the trap
was defined 'oil yield'. The difference between the
weight of charged substances and the residue plus oil
yields was defined 'gas yield' which contained the loss
during the experiment. The gas and oil yields were
usually less than 20% under the present conditions.
     The residual product in the reaction was ground

and stirred in 100 ml of quinoline for 1 hr at room
temperature, and filtered after centrifugation. This
extraction precedure was repeated until the filtrate
became colorless (usually 2~4 times). The quinoline
insoluble (QI), thus obtained, was washed with benzene
and acetone and then dried for weighing. The collected
filtrate was evaporated to dryness in vacuo and washed
with acetone for weighing. The degree of the solvolytic liquefaction was described with two ways of expre-
ssion, liquefaction yield (LY) and liquefying efficiency (LE), which were defined by equations (1) and (2),
respectively. QI and the coal fed in these equations
were moisture and ash free (maf) weights.

$$\text{Liquefaction Yield (LY) \% } = (1- \frac{QI^P - QI^S}{\text{coal fed}}) \times 100 \qquad (1)$$

$$\text{Liquefying Efficiency (LE) \% } = (1- \frac{QI^P - QI^S}{QI^C}) \times 100 \qquad (2)$$

where QI^P, QI^S, and QI^C are weights of quinoline insoluble in the residual product, in the original solvent,
and in the heat-treated coal at the liquefaction temperature without any solvent, respectively. LE may describe the increased yield of liquefaction by using the
solvent, indicating its efficiency for the liquefaction.

Results

Liquefaction of fusible coal at high temperature.
The liquefaction of Itmann coal, of which softening
point and maximum fluidity temperature are 417° and
465°C, respectively, was carried out at several temperatures using decacyclene as a liquefaction solvent.
The results are shown in Fig. 2, where the QI yield was
adopted as a measure of liquefaction extent. Because
the solubility of decacyclene in quinoline was rather
limitted, the QI contained a considerable amount of
decacyclene. Liquefaction of this coal proceeded
scarcely below 420°C of the softening temperature with
this solvent as well as pyrene. Above this temperature, the QI yield decreased sharply with the increasing liquefaction temperature until the resolidification
temperature of the coal. The maximum LY observed at
this temperature was estimated 67%, decacyclene being
assumed uncharged under the conditions. Above the resolidification temperature, the QI yield increased
sharply. The carbonization may start. Decacyclene was
known unreacted at 470°C in its single heat-treatment
(10), and in its cocarbonization with some coals(11),
although it is fusible. Cocarbonization of fusible

Table 1   Coals and their properties

Properties	coal	Itmann	West Kent.14	Taiheiyo
Proximate analysis (wt%)	ash	7.3	12.8	10.1
	volatile matter	19.9	53.0	45.9
	fixed carbon	72.8	34.2	37.7
ultimate analysis (wt%)	C	90.1	79.0	77.8
	H	4.6	5.1	6.0
	N	1.3	1.7	1.1
	S	0.5	4.6	0.2
	O(diff)	3.5	9.6	14.9
plasticity analysis	soften.temp.(°C)	417	387	
	max.fluid.temp.(°C)	465	425	non-
	max.fluid.(ddpm)	64	45	fusible
	final temp.(°C)	487	445	

Table 2   Coal liquefaction by pyrene derivatives
(reaction temp.=370°C, solvent/coal=3/1)

solvent	n*	coal	residue(%) QI	QS	distilate(%) oil	gas	L.Y.**	L.E.**
none	—	West Kent.	85.1	9.2	0.8	4.9	17	0
		Itmann	97.5	0.0	0.8	1.7	3	0
		Taiheiyo	85.4	0.0	2.2	12.4	16	0
pyrene	0	West Kent.***	19.9	69.9	6.0	4.2	24	8
		Itmann	20.4	75.2	3.1	1.3	23	18
		Taiheiyo	19.3	75.0	2.5	3.2	25	11
hydro-pyrene (No.1)	—	West Kent.	7.6	73.4	9.6	9.4	80	76
		Itmann	9.1	79.6	5.8	5.5	69	68
		Taiheiyo	6.3	85.0	4.5	4.2	83	80
hexyl-pyrene	0.83	West Kent.	19.3	51.3	12.8	16.5	26	11
		Itmann	20.9	63.1	8.5	7.5	17	15
		Taiheiyo	19.7	56.2	11.8	12.8	25	9
propyl-pyrene	0.85	West Kent.	14.4	66.3	3.5	16.1	49	38
		Itmann	14.9	66.3	7.2	11.6	43	42
		Taiheiyo	15.7	65.7	6.7	11.9	42	30
ethyl-pyrene	0.33	West Kent.	7.6	79.3	12.4	0.7	80	76
		Itmann	11.4	77.8	5.5	5.7	59	58
		Taiheiyo	13.4	76.8	2.9	6.9	52	42

* number of alkyl group introduced/one pyrene molecule
**L.Y.=Liquefaction Yield(%);   L.E.=Liquefying Effi-
ciency(%)   *** reaction temp.=390°C

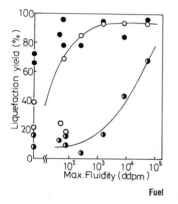

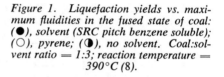

Fuel

Figure 1. Liquefaction yields vs. maximum fluidities in the fused state of coal: (●), solvent (SRC pitch benzene soluble); (○), pyrene; (◑), no solvent. Coal:solvent ratio = 1:3; reaction temperature = 390°C (8).

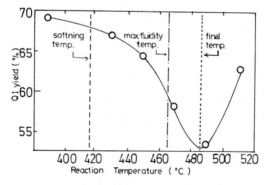

Figure 2.   Effect of reaction temperature on coal liquefaction yield. Coal = It-mann; solvent = decacyclene; solvent:coal ratio = 3:1.

coal with decacyclene developed a homogeneous optical
texture in the resultant coke, indicating their mutual
solubility (11).

Liquefaction of coals in alkylated and hydrogenated pyrenes. Table 2 shows liquefaction activity
of alkylated and hydrogenated pyrenes at 370°C, respectively. Although hexylpyrene showed just the same
liquefaction ability as pyrene, propylation and ethylation certainly improved the liquefaction activity of
pyrene with these coals of three different ranks. It
is of value to note that ethylpyrene showed LY of 80%
with West-Kentucky coal, which was significantly higher
than with other coals.
      Hydrogenation improved quite significantly the
liquefaction activity of pyrene with these coals. The
LY values with Taiheiyo and West-Kentucky coals reacted
to 80%, although the value with Itmann is rather low.
      The effect of hydrogenation extent on the lique-
faction activity was summarized in Table 3. As the

Table 3
Effect of hydrogenation extent on the liquefaction
activity (reaction temp.= 370 C, solvent/coal=3/1)

| | | | Itmann | | Taiheiyo | |
	solvent	n*	L.Y. (%)	L.E. (%)	L.Y. (%)	L.E. (%)
	No.1	2.2	69	68	83	80
hydro-	No.2	2.8	78	78	90	88
pyrene	No.3	4.3	72	71	85	82
	No.4	4.7	72	71	83	79

  L.Y.=Liquefaction Yield
  L.E.=Liquefying Efficiency
  * n =number of hydrogen atoms
        introduced/one pyrene molecule

number of hydrogen atoms introduced per one pyrene
molecule varied from 2.2 to 4.7 by using a variable
amount of lithium in Birch reduction. The liquefaction activity of hydrogenated pyrene was affected
slightly, reaching the maximum around three hydrogen
atoms per one molecule. It is obviously observed that
LY was always larger with Taiheiyo than Itmann.

Structural change of solvents and coals after the
liquefaction reaction. To analyse structural
change of solvents and coals in the liquefaction reaction, the solvent and coal should be separated.
Because the separation was rather difficult, it was

assumed that the benzene soluble and insoluble fractions after the liquefaction were derived from the solvent and coal, respectively. The procedure can be
shematically described in the following:

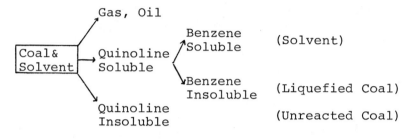

Fractionation scheme of liquefaction product

This assumption was verified by the following fact.
The amount of BS recovered in the liquefaction of Taiheiyo coal with pyrene was 73.3% as shown in Table 4.
This value corresponds to 97.7% of the starting amount
of pyrene and the BS fraction at the same time showed
the same NMR pattern to that of pyrene. This is also
true for hydropyrene, however the recovered BS was
rather low for ethylpyrene, indicating its reactivity.
      The recovery percentages of coal and solvent (BI/
coal fed, BS/solvent fed, respectively) calculated
based on the above assumption are summarized in Table
4. They were more than 85% except for the significantly low value for the coal recovery when hydropyrene
was used as the solvent. In the latter case, some
extent of the coal may be converted into the benzene
soluble. Nevertheless, analyses of BS and BI fractions
may inform the structural change of coal and solvent
after the liquefaction.
      Figures 3 and 4 show the NMR spectra of benzene
solubles after the liquefaction reaction using hydropyrene and ethylpyrene, respectively, together with
those of the solvents before liquefaction for comparison. The keys for identification of hydrogen observed
in the spectra are summarized in Fig. 5. The BS
derived from hydropyrene after the liquefaction lost
the resonance peaks at 2.5, 3.2 and 4.0 ppm extensively, although peaks at 2.0 and 2.8 ppm remained
unchanged as shown in Figure 3. In contrast, there
was essentially no change in the NMR spectra of ethylpyrene and its BS derivative as shown in Figure 4,
indicating that the BS derivative contained unchanged
ethylpyrene. However, the relatively low BS recovery
of this case suggests that the conversion of this

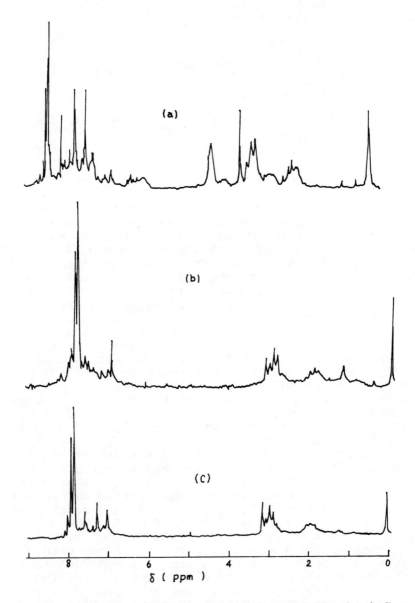

Figure 3.   NMR spectra of (a) hydropyrene, (b) its BS derivative after the liquefaction process, and (c) hydropyrene heated at 400°C

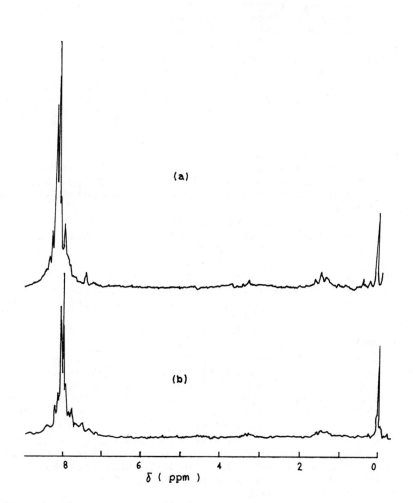

Figure 4.   NMR spectra of (a) ethylpyrene and (b) its BS derivative after the lique-faction. For liquefaction conditions, see Table 4.

compound into the carbonized matter may increase the BI
yield in comparison with other cases as shown in Table
4.

Table 5 shows the ultimate analysis of benzene
insoluble fractions after the liquefaction. H/C ratios
of these fractions were similar when no solvent or
pyrene was used, however when ethylpyrene and hydropyrene were used, the values were significantly low and
high, respectively. The hydrogenation of coal by the
hydrogen transfer from the hydrogen donating solvent
is strongly suggested in the latter case. The low H/C
value in the case of ethylpyrene may be explained in
terms of certain extent of carbonization, as suggested
by the low recovery of the solvent shown in Table 4.

Discussion

In a previous paper (5), the authors described the
liquefaction mechanism according to the properties of
the coal and the solvent. The coal was classified into
two categories.
   (1) fusible at the liquefaction temperature,
   (2) non-fusible at the liquefaction temperature.
The fusible coals can give a high liquefaction yield if
the high fluidity during the liquefaction is maintained
by the liquefaction solvent to prevent the carbonization. The properties of the solvent required for the
high yield with this kind of coal are miscibility, low
viscosity, radical quenching reactivity and thermal
stability not to be carbonized at the liquefaction
temperature as reported in literatures (12).

In contrast, the non-fusible coal requires the
solvation (extraction) or solvolytic reaction to be
liquefied. The solvation of non-polar organic compounds including the pitch may be rather limitted, so
that the solvolytic reaction is necessary for the high
liquefaction yield between the coal and the solvent.

The reaction may contain hydrogenation, alkylation, and depolymerization of coal molecules assisted
by the liquefying solvent. Through these reactions,
coal molecules can be converted to be fusible or
soluble in the solvent.

The present results are well understood by the
above mechanism. Itmann which is fusible at relatively
high temperature was not liquefied below 420°C with a
non-solvolytic solvent such as pyrene, however it was
significantly liquefied at 480°C of its maximum fluidity temperature in decacyclene of a stable aromatic
compound.

With solvolytic solvents, the fluidity of the coal

Table 4 Benzene extraction of solvolysis pitches
(coal=Taiheiyo, reaction temp.=370°C,solvent/coal=3/1)

solvent	residue		distilate		recovery*	
	BI (%)	BS (%)	oil (%)	gas (%)	coal (%)	solvent (%)
none	86.4	10.0	0.5	3.2	86.4	–
hydro-pyrene (No.2)	18.4	70.6	3.6	7.4	74.6	94.1
ethyl-pyrene	23.9	64.0	7.9	4.2	96.0	85.3
pyrene	22.4	73.3	0.9	3.4	89.6	97.7

$$* \text{ recovery} \quad \text{coal}(\%) = \frac{BI}{\text{coal fed}}$$

$$\text{solvent} = \frac{BS \text{ in residue}}{\text{solvent fed}}$$

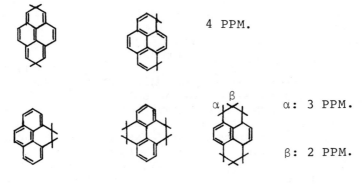

4 PPM.

α: 3 PPM.

β: 2 PPM.

Figure 5.   Proton identification

Table 5   Ultimate analysis of benzene
insoluble part of solvolysis pitch
(coal=Taiheiyo, reaction temp.=370°C,solvent/coal=3/1)

solvent	C(%)	H(%)	N(%)	others(%)	H/C
none	75.05	5.29	1.40	18.26	0.840
pyrene	76.25	5.43	1.53	16.75	0.849
ethyl-pyrene	82.24	4.93	1.80	11.03	0.714
hydro-pyrene	72.47	5.48	1.38	20.67	0.901

may not be a principal factor any more. Taiheiyo,
West-Kentucky, and Itmann coals of three different
ranks were sufficiently liquefied with hydropyrene
under atmospheric pressure at 370°C regardless of their
fusibility. The analyses of hydropyrene and the coal
before and after the liquefaction clearly indicate the
hydrogen transfer from the solvent to the coal substance. Lower rank coals look to show rather higher
reactivity in such liquefaction, probably because their
constituent molecule may have smaller condensed ring.

Transalkylation might be expected another kind of
the solvolitic reaction. However, the present results
suggest low probability with alkylated pyrenes as suggested by the NMR analyses. Instead, the increased
polarity by alkyl group or the enhanced reactivity of
the carbonization precursor from alkylpyrene, especially ethylpyrene, may be responsible for a considerable liquefaction yield. The recovery of the solvent
becomes difficult by its latter conversion as observed
in the present study.

Failure of hexylpyrene as the liquefaction solvent
may be due to the easy dealkylation (13) or high carbonization reactivity probably catalyzed by coals.
Transalkylation for coal-liquefaction may require the
acid-catalyst (14) or high pressure (15).

Details on the liquefaction of slightly fusible
coals (West-Kentucky) are not discussed in the present
study. Rather huge storage of such coals may be most
suitable for further development. At moment, moderate
hydrogenation is extensively investigated (4), however
the authors would propose the possibility of their
extractive liquefaction by using the proper solvent or
improving their fusibility. High liquefaction yield
by ethylpyrene for this coal is suggestive. Such
liquefaction may be economically interesting. Further
study will be published soon (16).

Acknowledgement

This research was partly supported by the Ministry
of Trade and Industry through the Sun-Shine project and
partly by specified research project on energy adminis-
trated by the Ministry of Education.

Abstract

Atmospheric liquefaction of coals of three ranks
were studied at temperature range of 370~470°C, using
decacyclene, pyrene, alkylated pyrene and hydrogenated
pyrene as the liquefying solvent. Coal of high

softening temperature was liquefied using decacyclene, which is stable at the temperature. Hydropyrene gave high liquefaction yields regardless of the coal rank. Significant activity of ethylpyrene is of value to be noted.

The liquefaction mechanism was discussed by distinguishing the fusible coal from non-fusible one. The importance of solvolytic hydrogen transfer is pointed for the liquefaction of non-fusible coal under atmospheric pressure.

Literature Cited

(1) (a) Yavorsky, P. M. J. Fuel Inst. Japan, 1976, 55, 143.
     (b) Van Kravelen, D. W. "Coal" Elsevier Amsterdam, 1961.
     (c) Oblad, A. G. Catal. Rev., 1976, 14 (1), 83.
     (d) Ellington, R. T. "Liquid Fuels from Coal", Academic Press Inc. 1977.
(2) Kamiya, Y. J. Fuel Inst. Japan, 1977, 56, 319.
(3) Bartle, K. D.;Martin, T. G.;Williams, D. F. Fuel, 1975, 54, 226, 1979, 58, 413.
(4) Pott, A.;Brock, H. Fuel, 1934, 13, 91, 125, 154., Ruberto, R. G.;Cronaver, D. C.;Jewell, D. M.; Seshadri K. S. Fuel, 1977, 56, 25.
(5) Neavall, R. C. Fuel, 1976, 55, 237.
(6) Yamada, Y.;Honda, H.;Oi, S.;Tsutsui, H. J. Fuel Inst. Japan, 1974, 53, 1052.
(7) Osafune, K.;Arita, S.;Yamada, Y.;Kakiyama H.; Honda, H.;Tagawa, N. J. Fuel Inst. Japan, 1976, 55, 173.
(8) Mochida, I.:Takarabe, A.;Takeshita, K. Fuel, 1979, 58, 17.
(9) Reggel, L.;Friedel, R. A.;Wender, I. J. Org. Chem., 1975, 22, 891, Mochida, I.;Kudo, K.; Takeshita, K. Fuel, 1974, 53, 253, 1976, 55, 70.
(10) Mochida, I.;Miyasaka, H.;Fujitsu, H.;Takeshita, K. Carbon(Tanso), 1977, No. 92, 7. Mochida, I.; Marsh, H. Fuel, 1979, 58, 626.
(11) Mochida, I.;Grint, A.;Marsh, H. Fuel, 1979, 58, 633.
(12) Scarah, W. P.;Dilon, R. R. Ind. Eng. Chem., Process Design Dev., 1975, 15, 122, Sato, H.; Kamiya, Y., J. Fuel Inst. Japan, 1977, 56, 103.
(13) Mochida, I.;Maeda, K.;Takeshita, K. Fuel, 1976, 55, 70.
(14) Schlosberg, R. H.;Gorbaty, R. H.;Aczel, T. J. Am. Chem. Soc., 1978, 100, 4188.
(15) Shibata, M.;Arita, S.;Honda, H. 15th Coal Science

Meeting, JAPAN Abstract, 1978, p.26.
(16) Mochida, I.;Tahara, T.;Iwamoto, K.;Fujitsu, H.;
Korai, Y.;Takeshita, K. Nippon Kagakukai Shi (J.
Chem. Soc. Japan) in press (May, 1980).

RECEIVED March 28, 1980.

The Effect of Catalyst Concentration, Temperature, and Residence Time on the Chemical Structure of Coal Hydropyrolysis Oils

JOHN R. KERSHAW, GORDON BARRASS, and DAVID GRAY

Fuel Research Institute of South Africa, P.O. Box 217, Pretoria, South Africa 0001

Three of the most important parameters that affect the hydrogenation of coal and the products obtained are the catalyst, the temperature and the residence time. The effects of these parameters on the chemical nature of the products are still not fully understood.

In this paper we have looked firstly at the effect that the catalyst concentration, secondly at the effect that the reactor temperature and finally at the effect that the residence time at temperature have on the chemical structure of the oils (hexane soluble product) produced on hydropyrolysis (dry hydrogenation) of a high volatile bituminous coal. Generally, the hydropyrolysis conditions used in this study resulted in oil yields that were considerably higher than the asphaltene yields and this study has been limited to the effects that the three reaction conditions have on the chemical nature of the oils produced.

For the study of catalyst concentration tin as stannous chloride was used as the catalyst and the concentration range studied was 0 - 15% by weight of the coal. Stannous chloride is one of the best, if not the best, catalyst for conversion of coal to liquid products. For the study of the effect of the reactor temperature, the temperature range studied was from 400 - 700°C.

In the work reported here hydrogenation of coal was carried out in the absence of any vehicle oil in a semi-continuous reactor which allowed the volatile product to be swept from the reactor by a continuous stream of hydrogen. The great majority of coal hydrogenation oils studied by other workers were produced using a liquid vehicle. The vehicle oil may have an effect on the chemical reactions taking place and, therefore, on the chemical composition of the product. For example, it has been suggested (1) that the main role of the catalyst in coal hydrogenation where a solvent is used is to hydrogenate the solvent.

0–8412–0587–6/80/47–139–273$05.00/0

EXPERIMENTAL

<u>Materials</u> The coal (0.50 to 0.25 mm fraction) used was
from the New Wakefield Colliery, Transvaal. Analysis, air dried
basis: Moisture 4.9; Ash 14.9; Volatile Matter 32.8%;   dry
ash-free basis: C, 79.2; H, 5.4; N, 2.1; S, 2.3%.
     The catalyst was analytical grade stannous chloride.
Stannous chloride was dissolved in water and added to the coal
as an aequous solution. The resultant slurry was mixed by
stirring and then dried under nitrogen.

<u>Hydrogenation</u> Hydrogenation was carried out in a reactor
similar to the "hot-rod" reactor (2, 3) designed by Hiteshue
et al. The coal (25 g) impregnated with catalyst was mixed with
sand to limit agglomeration (4) and is held within the reactor
with steel wool plugs. The reactor was heated by direct resistance heating at a rate of 200°C/minute. Hydrogen, at a flow
rate of 22 ℓ/minute and a pressure of 25 MPa was continuously
passed through the fixed bed of coal/catalyst/sand. The volatile
products were condensed in a high-pressure cold trap. For the
study of the effect of the catalyst concentration and the temperature, the time at temperature was 15 minutes. The other conditions are given in the Results and Discussion section.
     The product was removed from the cooled reactor and from the
condenser with the aid of toluene. The solid residue was extracted with boiling toluene (250 ml) in a soxhlet extractor for 12
hours. The toluene solutions were combined and the toluene
removed under reduced pressure. Hexane (250 ml) was added to the
extract and it was allowed to stand for 24 hours with occasional
shaking. The solution was filtered to leave a residue (asphaltene) and the hexane was removed from the filtrate under reduced
pressure to give the oil.

<u>Fractionation of the oils</u> The oils were fractionated by
adsorption chromatography on silica gel. The column was eluted
successively with 40 - 60°C petroleum ether (12 fractions),
40 - 60 petroleum ether/toluene (increasing proportions of
toluene, 5 fractions), toluene, chloroform and methanol. The
oils were also extracted with 2 M NaOH and 2 M HCl to give the
acids and bases, respectively.

<u>Analyses</u> I.r. spectra were measured as smears on sodium
chloride plates or as a solution in carbon tetrachloride using
a Perkin-Elmer 567 grating spectrophotometer, while u.v. spectra
were measured as a solution in hexane (spectroscopic grade) using
a Unicam SP 1700 instrument. Fluorescence and phosphorescence
spectra were recorded as described elsewhere (5, 6).
     [1]H n.m.r. spectra were recorded for the oils in deutero-
chloroform or carbon tetrachloride at 90 MHz with tetramethy-
silane as an internal standard using a Varian EM 390 instrument.

Broadband proton-decoupled pulse Fourier transform [13]C n.m.r.
were recorded in deuterochloroform at 20 MHz using a Varian
CFT-20 spectrometer.

Molecular weights were determined by vapour pressure osmometry in benzene solution using a Knauer apparatus. 5 concentrations over the range 1 - 5 g/ℓ were employed and the molecular
weight was obtained by extrapolation to infinite dilution.

The viscosities of the oils were measured using a Haake
Rotovisco RV3 viscometer with a cone and plate sensor at 20°C.

Elemental analysis of the oils was carried out as follows:
carbon and hydrogen by micro combustion using a Perkin-Elmer 240
Elemental Analyzer; sulphur by X-ray fluorescence using a
Telsec Lab X-100 apparatus; nitrogen by chemiluminescence using
a Dohrmann DN-10 apparatus.

RESULTS AND DISCUSSION

The Effect of Catalyst Concentration   The first parameter
that was studied was the effect of the catalyst concentration.
Samples impregnated with 1, 5, 10 and 15% tin as stannous
chloride and a sample with no catalyst were hydrogenated at 450°C
to investigate the effect that increasing catalyst concentration
has on the composition of the oil (hexane soluble portion) formed.

From elution chromatography, the percentage of aliphatic
hydrocarbons, aromatic hydrocarbons and polar compounds were
obtained. The percentage of polar compounds in the oil decreased
as the catalyst concentration increased (see Figure 1) with
mainly an increase in the percentage of aromatic hydrocarbons.
There was also a decrease in the percentage of both acids and
bases in the oil as the catalyst concentration increases as
shown in Figure 1.

The amount of sulphur and nitrogen in the oils decreases
with catalyst concentration. The decrease in sulphur content is
shown in Figure 2 while the nitrogen content of the oils decreased from 1.3% to 0,9% when the catalyst concentration was increased from 1% to 10%.

Thus the results from elution chromatography, extraction
with acid and base and elemental analysis show that as the
catalyst concentration used increases, the heteroatom content
of the resultant oil decreases.

The fractions from elution chromatography were studied by
a number of spectroscopic methods, [1]H n.m.r., i.r., u.v., fluorescence and phosphorescence spectroscopy. Equivalent fractions
from chromatographic separation of the various oils showed no
significant differences in their spectra and it appears that the
composition of the fractions was independent of the catalyst
concentration used to produce the oil. Though, as previously
mentioned the amounts of the various fractions especially the
polar fractions differ with the catalyst concentration. G.l.c.
analysis of the saturate fractions also indicated no changes
with different catalyst concentrations.

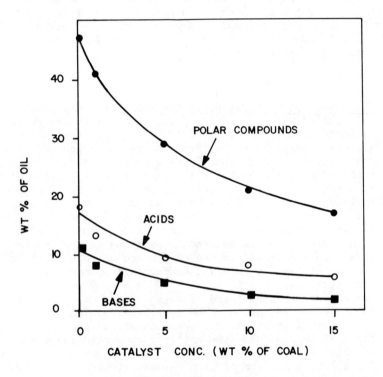

Figure 1.   Variation of polar compounds, acids, and bases with catalyst concentration

[1]H n.m.r. spectra were recorded for the whole oils, the acids, bases and neutral components. The percentage of hydrogens in aromatic, benzylic and aliphatic environments showed no change with catalyst concentrations. [13]C n.m.r. spectra were also recorded for the oils produced. No discernible differences could be found between the spectra of the oils. A more detailed description of the spectra of these oils is reported elsewhere (6).

The spectroscopic evidence indicates that the catalyst concentration had very little effect on the "gross" hydrocarbon structure present and this is substantiated by the [H]/C atomic ratios of the oils which showed no significant change with catalyst concentration.

It was obvious on visual examination of the oils that the greater the catalyst concentration used, the less viscous was the oil produced. The decrease in viscosity with catalyst concentration is shown in Figure 3.

It has been reported that the molecular weight of coal liquids affects the viscosity (7, 8). However, the decrease in molecular weight that occurred with increasing catalyst concentration was relatively small, as shown in Figure 4. We feel that this relatively small change in molecular weight would not cause such a noticeable change in viscosity unless changes in the chemical nature of the oil also contributed to the viscosity reduction.

Sternberg et al. (7) showed that the presence of asphaltenes in coal-derived oils caused a marked increase in the viscosity. This group also showed that these asphaltenes were acid-base complexes and that hydrogen bonding occurs between the acidic and basic components of asphaltenes (9, 10). Recent work (8, 11) on coal liquefaction bottoms has shown the importance of hydrogen bonding on the viscosity of coal liquids.

The reduction of polar compounds in the oil with increasing catalyst concentration could reduce hydrogen bonding and, therefore, the viscosity of the oil. To further look at this possibility, i.r. spectra were recorded at the same concentration in CCl_4 for each of the oils (see Figure 5). The i.r. spectra showed sharp peaks at 3610 cm^{-1} (free OH), 3550 cm^{-1} (2nd free OH) and 3480 cm^{-1} (N-H) and a broad peak at ca 3380 cm^{-1} which is assigned to hydrogen bonded OH. This band decreases with increasing catalyst concentration (see Figure 5) indicating that hydrogen bonding in the oil decreases with increasing catalyst concentration used to produce the oil. The band at ca 3380 cm^{-1} was shown to be due to intermolecular hydrogen bonding (12, 13) by recording the spectrum of a more dilute solution (using a longer path length cell), the 3380 cm^{-1} band diminished with an increase in the 3610 cm^{-1} peak.

Though, as previously stated, the [1]H n.m.r. spectra of the oils were very similar, there was one noticeable difference with catalyst concentration and that was in the position of the OH peak. As the catalyst concentration used to produce the oil increases then the OH signal is shifted to higher field (see

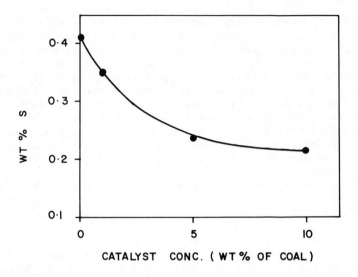

Figure 2.   *Variation in sulfur content with catalyst concentration*

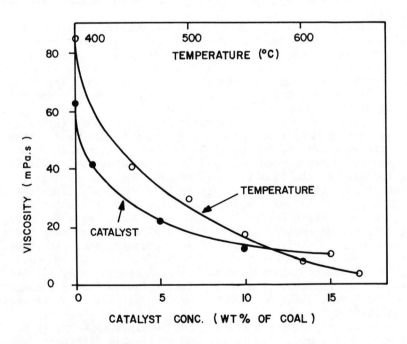

Figure 3.   *Variation in viscosity with catalyst concentration and reactor temperature*

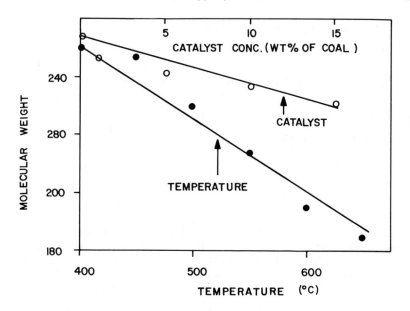

Figure 4. Variation in molecular weight with catalyst concentration and reactor temperature

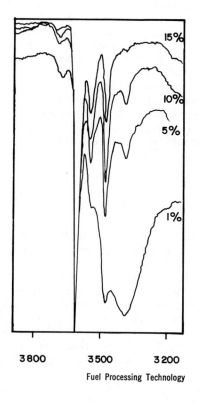

Figure 5. Partial IR spectra of oils (various catalyst concentrations) (6)

Table I) indicating less hydrogen bonding.

Table I

^{1}H N.m.r. shifts of OH Signal

Catalyst Concentration (% Sn)	0	1	5	10	15
δ OH (0.2 g of oil in 1 ml of CCl_4)	5.7	5.2	4.7	4.4	4.1

It would appear that increasing the amount of stannous chloride catalyst, under our experimental conditions, as well as increasing the amount of oil formed decreases the amount of polar compounds in the oil which decreases the hydrogen bonding and therefore helps to decrease the viscosity of the oil. Spectroscopic evidence indicates that there is little change in the hydrocarbon structures present.

The Effect of Temperature   The second parameter that we looked at was the effect of temperature. The temperature range studied was from 400°C to 700°C and tin (1% of the coal) as stannous chloride was used as the catalyst.

^{1}H n.m.r. spectra of the oils were recorded for the range of temperatures and the protons were assigned as aromatic, phenolic OH, benzylic and aliphatic. There was an increase in the percentage of aromatic protons and a decrease in the percentage of aliphatic protons as the temperature increases, while the percentage of benzylic protons remained constant (see Figure 6). It, therefore, appears that as the hydrogenation temperature increases side groups are lost and that the C-C bond directly attached to the aromatic ring is more stable than those further from the ring. The molecular weight of the oil decreases with temperature (see Figure 4) as would be expected if side chains are being removed.

The aromaticity of the oils, as calculated from the ^{1}H n.m.r. data using the Brown-Ladner equation (14), increase with temperature (see Figure 7) as does the $^C/_H$ atomic ratio as shown in Figure 7.

^{13}C n.m.r. spectra were recorded for the oils produced at 400°, 450°, 550° and 600°C. As the temperature increased the aromatic carbon bands became much more intense compared to the aliphatic carbon bands (see Figure 8). Quantitative estimation of the peak areas was not attempted due to the effect of variations in spin-lattice relaxation times and nuclear Overhauser enhancement with different carbon atoms. Superimposed on the aliphatic carbon bands were sharp lines at 14, 23, 32, 29, and 29.5 ppm, which are due to the α, β, γ, δ, and ε-carbons of long aliphatic chains (15). As the temperature increases, these lines

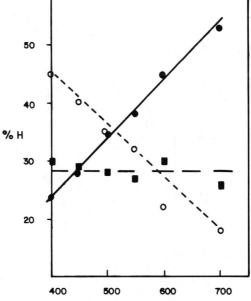

% H

TEMPERATURE (°C)

Fuel Processing Technology

Figure 6. Variation of hydrogen distribution with reactor temperature (6): (●), aromatic (δ 8.7 − 6.1 ppm); (○), aliphatic (δ < 1.9 ppm); (■), benzylic (δ 3.8 − 1.9 ppm).

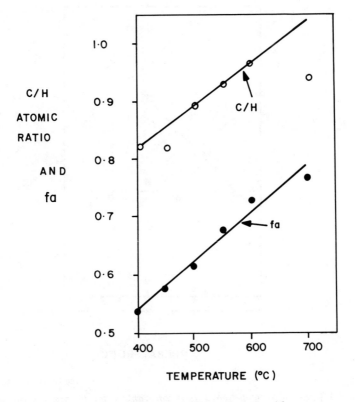

Figure 7. Variation of aromaticity and C/H atomic ratio with reactor temperature

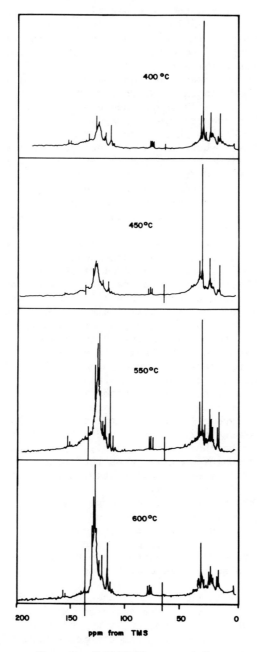

Figure 8. C-13 NMR spectra of oils

become smaller compared to the other aliphatic bands and this
is especially noticeable in the spectrum of the 600°C oil. The
ε line was approximately four times the intensity of the α and β
lines at 400°C and 450°C, at 550°C approximately three times and
at 600°C only about twice the intensity. It would seem that as
the temperature increases, the long aliphatic chains are reduced
in both number and length. (G.1.c. analysis of the saturate
fractions from elution chromatographic separation showed that as
the hydrogenation temperature increases there was a decrease in
the percentage of the higher alkanes and an increase in the
percentage of their shorter chained analogues in the saturate
hydrocarbon fractions). It was also noticeable when comparing
the spectrum of the 600°C oil to the spectra of the 400°C and
450°C oils that the intensity of bands due to CH_3 α to aromatic
rings (19 - 23 ppm from TMS (16)) had increased in intensity
compared to the other aliphatic bands. This agrees with the [1]H
n.m.r. results which showed no change in the percentage of benzylic protons while the percentage of aliphatic protons decreased.

It has also been recently independently reported (17, 18)
that there is an increase in the aromaticity of the liquid product
with increasing reactor temperature, but in both these reports
the temperature range studied was very limited. Whitehurst, et al.
in their studies of the solvent refining of coal ascribe the
increased aromatic nature of the product obtained at 450°C
compared to 425°C to mainly increased internal rearrangement and
aromatization rather than dealkylation of aromatic rings occuring
at the higher temperature. However, the considerable increase
in the gas yield, the decrease in molecular weight, the decrease
in the percentage of aliphatic protons while the percentage of
benzylic protons remained unchanged and the decrease in the
number of and length of long aliphatic chains with increasing
reactor temperature indicate that, under the hydropyrolysis
conditions described here, it is mainly the cleavage of aliphatic
side groups and chains to give gases that accounts for the
increased aromaticity of the oil rather than dehydrogenation of
naphthenes and hydroaromatics.

Elution chromatography gave the percentage of aliphatic
hydrocarbons, aromatic hydrocarbons and polar compounds in the
oil. There was a reduction in the percentage of polar compounds
in the oil (see Figure 9) with subsequent increase in the
aromatic percentage as the reactor temperature increased. The
sulphur content also decreased with temperature as shown in
Figure 9.

The viscosity of the oil decreases considerably with temperature as shown in Figure 3. The decrease in viscosity is expected
as there was a decrease in molecular weight with increasing
temperature and also a decrease in the percentage of polar
compounds in the oil.

The boiling range of the oils, as indicated by thermal
gravimetric analysis and simulated distillation shows a marked
improvement as the reactor temperature increases.

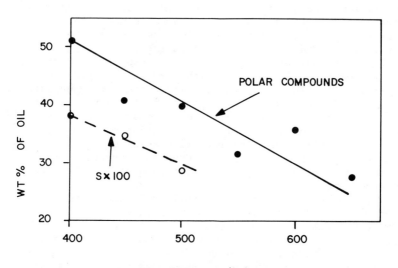

Figure 9.   Variation of polar compounds and sulfur content with reactor temperature

The effect of increasing the hydrogenation temperature, under the conditions used here, is to give a "lighter", more aromatic oil of lower molecular weight containing fewer long aliphatic chains. The viscosity of the oil and the percentage of polar compounds in the oil also decrease with increasing temperature.

The Effect of Residence Time The final parameter that was studied was the solid residence time. In the semi-continuous reactor used for this study, the volatile product is swept from the reactor by a continuous stream of hydrogen and, therefore, there is both a vapour and solid residence time. It is this latter parameter that has been studied here and the solid residence time was considered to be the time that the reactor spends at temperature. For the study of the residence time, tin (1% of the coal) as stannous chloride was used as the catalyst and the other conditions are given in Table II.

Compositional analysis shows a decrease in the percentage of polar compounds in the oils with increasing residence time (see Table II). The decrease in polar content is substantiated by a lower sulphur content and results in a lower viscosity (see Table II). The oil becomes more aromatic, as shown by [1]H n.m.r. spectroscopy (see Table II), with increasing time at temperature, while the molecular weights showed little change. G.l.c. analysis of the saturate hydrocarbon fractions from elution chromatography indicated little change in the saturates with residence time.

The changes that occur with solid residence time in the "hot-rod" reactor were not very pronounced because only the non-volatile portion of the oil remaining on the coal bed would be expected to undergo secondary reactions such as aromatization and loss of heteroatoms. However, the oils from the "hot-rod" reactor were also compared with those obtained in a rotating autoclave with much longer solid and vapour residence times and the changes with residence time were more noticeable in this case as can be seen in Table II.

CONCLUSIONS

The results reported here have important implications for coal hydrogenation processes.

Increasing the catalyst concentration appears to have a wholly beneficial effect on the quality of the oil obtained. The heteroatom content and viscosity are reduced while the gross hydrocarbon structure is little changed. However, for economic reasons high catalyst concentrations are unlikely to be used.

Increasing the reactor temperature has both positive and negative effects on the quality of the oil. The beneficial effects are that a "lighter" product of lower molecular weight and viscosity, and containing fewer polar compounds is formed as the reactor temperature increases. However, the increase in

Table II

The Effect of Residence Time on the Composition of the Oil Produced

Temperature (°C)	Residence Time (min)	Hydrogen Type Aromatic/Phenolic OH/Benzylic/Aliphatic (% of total hydrogens)				Polar Compounds (Wt%)	S (Wt%)	Viscosity (mPa.s)	Mol. Wt[+]
500	2½	29	2	28	41	48	0.36	97	234
500	5	33	3	29	35	42	–	–	–
500	15	35	2	28	35	40	0.29	29	230
500	25	35	2	29	34	38	–	–	–
450	15	28	3	29	40	41	0.35	41	–
450	60*	38	3	31	28	27	0.17	19	–

*Using Rotating Autoclave

[+]Determined by vapour pressure osmometry

aromaticity with temperature and the consequential decrease in $^H/C$ atomic ratio are detrimental especially if it is desired to produce diesel and jet fuel.

Increasing the residence time also has beneficial and detrimental features. On the positive side the heteroatom content is reduced while on the negative side the aromaticity increases.

LITERATURE CITED

1. Ruberto, R.G., Cronauer, D.C., Jewell, D.M. and Seshadri, K.S. Fuel, 1977, 56, 25.
2. Hiteshue, R.W., Anderson, R.B. and Schlessinger, M.D. Ind. Eng. Chem., 1957, 49, 2008.
3. Gray, J.A., Donatelli, P.J. and Yavorsky, P.M. Preprints, Amer. Chem. Soc., Div. of Fuel Chem., 1975 20 (4), 103.
4. Gray, D. Fuel, 1978, 57, 213.
5. Kershaw, J.R. Fuel, 1978, 57, 299.
6. Kershaw, J.R., Barrass, G. and Gray, D. Fuel Process. Technol., in press.
7. Sternberg, H.W., Raymond, R. and Schweighardt, F.K. Preprints, Amer. Chem. Soc., Div. of Petrol. Chem., Chicago Meeting, Aug. 24-29, 1975.
8. Schiller, J.E., Farnum, B.W. and Sondreal, E.A. Preprints, Amer. Chem. Soc., Div. of Fuel Chem., 1977, 22 (6), 33.
9. Sternberg, H.W., Raymond, R. and Schweighardt, F.K. Science, 1975, 188, 49.
10. Schwieghardt, F.K., Friedel, R.A. and Retcofsky, H.L. Appl. Spectrosc., 1976, 30, 291.
11. Gould, K.A., Gorbaty, M.L. and Miller, J.D. Fuel, 1978, 57, 510.
12. Bellamy, L.J., "The Infrared Spectra of Complex Molecules", 3rd edn., Chapman and Hall, London, 1975.
13. Brown, F.K., Friedman, S., Makovsky, L.E. and Schweighardt, F.K. Appl. Spectrosc., 1977, 31. 241.
14. Brown, J.K. and Ladner, W.R. Fuel, 1960, 39, 87.
15. Pugmire, R.J., Grant, D.M., Zilm, K.W., Anderson, L.L., Oblad, A.G. and Wood, R.E. Fuel, 1977, 56, 295.
16. Bartle, K.D., Martin, T.G. and Williams, D.F. Chem. Ind., 1975, 313.
17. Whitehurst, D.D., Farcasiu, M. and Mitchell, T.O. "The Nature and Origan of Asphaltenes in Processed Coals", EPRI Report AF-252, Project RP-410-1, Annual Report, February 1976.
18. Ouchi, K., Chicada, T. and Itah, H. Fuel, 1979, 58, 37.

RECEIVED March 28, 1980.

LIQUEFACTION MECHANISMS

Thermal Treatment of Coal-Related Aromatic Ethers in Tetralin Solution

Y. KAMIYA, T. YAO, and S. OIKAWA

Faculty of Engineering, University of Tokyo, Hongo, Tokyo 113, Japan

The important elementary reactions of coal liquefaction are the decomposition of coal structure with low bond dissociation energy, the stabilization of fragments by the solvent and the dissolution of coal units into the solution.

These reactions proceed smoothly in the presence of hydrogen donating aromatic solvent (1-4) at temperatures from 400°C to 450°C, resulting in the formation of so called solvent refined coal with carbon content of 86-88% on maf basis independent of coalification grade of feed coal.

While, oxygen containing structures of coal must be playing important parts in the course of coal liquefaction. It will be key points that what kinds of oxygen containing structure are decomposed and what kinds of structure are formed in the course of reaction. It has been proposed (5,6) and recently stressed (7-11) that the units of coal structure are linked by ether linkage.

We have studied the thermal decomposition of diaryl ether in detail, since the cleavage of ether linkage must be one of the most responsible reactions for coal liquefaction among the various types of decomposition reaction and we found that the C-O bond of polynucleus aromatic ethers is cleaved considerably at coal liquefaction temperature.

Experimental

Tetralin and 1-methylnaphthalene were reagent grade and were used after washing with sulfuric acid, alkali, and water and the subsequent distillation at 70°C under reduced pressure. Various additives and model compounds were reagent grade, and some of them were used after recrystallization. Phenyl naphthyl ether and phenyl 9-phenanthryl ether were synthesized by refluxing a mixture of aryl bromide, phenol, Cu_2O and γ-collidine (12).

0–8412–0587–6/80/47–139–291$05.00/0

Samples were added to 300 ml or 90 ml magnetic stirring (500 rpm) autoclaves. After pressurizing with hydrogen, the autoclave was heated to the reaction temperature within 45 min and maintained at the temperature for the desired reaction time.

At the completion of a run, the autoclave was cooled by electric fan to room temperature and the autoclave gases were vented through gas meter and analyzed by gas chromatography. Liquid portions of the samples were subjected to gas chromatographic analysis to determine the composition of products.

Results and Discussion

Thermal Treatment of Various Aromatic Compounds. In order to study the reaction of coal structure, various aromatic compounds were chosen as the coal model and treated at 450°C. The conversion of the reaction along with the detected products were shown in Table I.

Recently, the thermal decomposition of diaryl alkanes such as dibenzyl and 1,3-diphenylpropane has been studied by Sato and coworkers (13), Collins and coworkers (14). These compounds were confirmed to be decomposed to alkylbenzenes gradually as a function of carbon chain length.

Although diphenyl ether and dibenzofuran were very stable at 450°C, 2,2'-dinaphthyl ether was decomposed slowly and benzyl ethers completely.

The apparent activation energy for the thermal decomposition of phenyl benzyl ether was calculated to be 50 kcal/mole, since the first order rate constants were 1.39×10^{-4} at 320°C, 5.19×10^{-4} at 340°C and 9.52×10^{-4} S^{-1} at 350°C, respectively.

These results imply that highly aromatic ether linkages will be considerably broken at coal liquefaction temperatures resulting in a main source of phenolic groups of the dissolved coal.

Phenolic compounds were confirmed to be very stable against thermal treatment. Diphenyl methanol and benzophenone were stable against decomposition but hydrogenated to form diphenylmethane quantitatively. Phenyl benzyl ketone was found to be partially hydrogenated or decarbonylated to form diphenyl alkanes.

Naphthoquinone was completely eliminated and hydrogenated to naphthol and dihydroxynaphthalene as reported by Brower (15).

Carboxylic acid and carboxylate were completely decarboxylated to the parent hydrocarbons. According to Brower, carboxylic acid is quite stable in glass apparatus, but decomposed completely in a stainless steel autoclave.

The Thermal Decomposition of Aromatic Ethers. According to the results of Table I, the bond scission of oxygen containing polynucleus aromatic structure of coal at liquefaction temperature of 450°C seems to occur mainly at methylene or ether structures. Therefore, it will be very important to study the

TABLE I THERMAL TREATMENT OF COAL-RELATED AROMATIC COMPOUNDS IN THE PRESENCE OF TETRALIN (Tetralin 220 mmole, 1-Methylnaphthalene 140 mmole, Model compounds 10 mmole)

Model compounds	Reaction conditions Temp. (°C)	Time (min)	Conversion(%) Model compounds	Products (mole% to reacted model compounds)
Diphenylmethane	450	30	1.7	Benzene 100, Toluene 95
Dibenzyl	450	30	31.1	Toluene 200
Dibenzyl	450	60	54.1	Toluene 194
Diphenyl ether	450	30	0	----------
Dibenzofuran	450	30	3.3	----------
Phenyl benzyl ether	400	30	100	see Table II
Dibenzyl ether	400	30	65	see Table II
2,2'-Dinaphthyl ether	450	120	23.3	see Table II
Phenyl 1-naphthyl ether	440	120	25	see Table II
Phenyl 9-phenanthryl ether	440	120	45.5	see Table II
Phenyl 9-phenanthryl ether	450	120	64.8	see Table II
Benzyl benzoate	450	30	100	Benzene 53, Toluene 98
Benzophenone	450	30	29.4	Diphenylmethane 100
Benzyl phenyl ketone	450	30	25.3	Dibenzyl 37, Diphenylmethane 37, Benzene 5, Toluene 7
1-Naphthol	450	30	1.4	----------
2-Naphthol	450	30	1.4	----------
Diphenylmethanol	450	30	79	Diphenylmethane 96
1,4-Naphthoquinone	450	30	100	1-Naphthol 74, Naphthalene
2-Naphthoic acid	450	30	100	Naphthalene
Diphenyl amine	450	30	8.2	Benzene 60, Aniline(trace)
Diphenyl sulfide	450	30	10.7	Benzene 183

charcteristics of these structures in the thermolysis.

It has been often proposed that the units of coal structure are linked by ether linkages. Recently, Ruberto and his co-workers (7,8), Ignasiak and Gawlak (9) concluded that a signifi-cant portion of the oxygen in coal occurs in ether functional groups.

Thermal decomposition of seven diaryl ethers at various reaction conditions and the composition of reaction products are shown in Table II.

TABLE II   THERMAL DECOMPOSITION OF DIARYL ETHERS IN TETRALIN

Diaryl ether	Temp. (°C)	Time (min)	Conver-sion(%)	Products(mole% of reacted ether)
Diphenyl ether	450	30	0	-----
	450	120	2	-----
Dibenzofuran	450	30	3.3	-----
Phenyl 1-naphthyl ether	440	120	25	PhOH 66
2,2'-Dinaphthyl ether	450	60	12.5	2-Naphthol 84
	450	120	23.3	2-Naphthol 63
Phenyl 9-phenanthryl ether	440	120	45.5	PhOH 50, Phenanthrene 15
	450	120	64.8	PhOH 46, Phenanthrene 17
Phenyl benzyl ether	320	30	31.4	$PhCH_3$ 55, PhOH 50, Benzylphenol 40
	400	30	100	$PhCH_3$ 61, PhOH 66, Benzylphenol 27
Dibenzyl ether	400	30	65	$PhCH_3$ 106, PhCHO 73, PhH 12

The decompositions of phenyl benzyl ether and dibenzyl ether proceeded very rapidly at 400°C, and these results corre-sponded well to the low bond dissociation energy of $PhCH_2$-O.

Phenyl benzyl ether was mostly converted to toluene and phenol, but partly isomerized to benzyl phenol.

Although diphenyl ether and dibenzofuran were very stable for thermolysis at 450°C for 120 min, the rate of decomposition increased with increasing the number of benzene nucleus, that is, 2,2'-dinaphthyl ether was converted to the value of 23.3% and phenyl phenanthryl ether 64.8% at the same reaction conditions.

Thus, diaryl ethers with polynucleus aromatic rings were found to be decomposed considerably fast at coal liquefaction conditions.

The high yield of toluene over 100% was confirmed to be
due to the hydrogenation of benzaldehyde by tetralin. As shown
in Figure 1, the yield of toluene increases with increasing
reaction time, on the other hand, benzaldehyde decreases
gradually after reaching a maximum value, giving toluene as the
hydrogenated product and benzene and carbon monoxide as the
decomposition products.

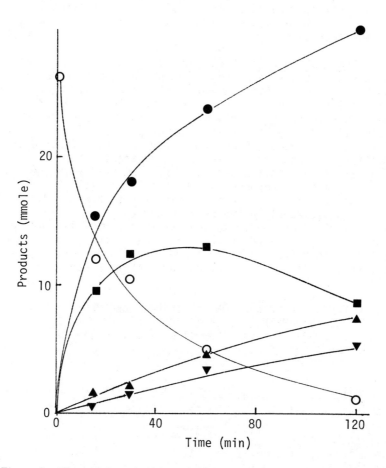

Figure 1. Thermal decomposition of dibenzyl ether (26 mmol) in Tetralin at
400°C: (○), PhCH₂OCH₂Ph; (●) PhCH₃; (■), PhCHO; (▲), PhH; (▼), CO.

Dibenzyl ether was entirely converted to toluene and benzaldehyde. The formation of products can be explained by the following reaction shceme.

$$PhCH_2OCH_2Ph \longrightarrow PhCH_2O\cdot + \cdot CH_2Ph \longrightarrow PhCHO + PhCH_3$$

$$PhCHO \longrightarrow PhH + CO \qquad PhCHO + 2H_2 \longrightarrow PhCH_3 + H_2O$$

These results strongly suggest that in coal structure the covalent bond of benzyl ethers composed of aliphatic carbon and oxygen will be entirely cleaved at temperatures lower than 400°C, and the covalent bond composed of aromatic carbon and oxygen will be considerably decomposed at 450°C, since the unit structure of bituminous coal is considered to be composed of polynucleus of several benzene rings.

The Effect of Phenolic Compounds on the Decomposition of Diaryl Ethers. It has been known that phenolic compounds (16, 17) in the presence of hydrogen-donating solvent have a remarkable effect on the dissolution of coal. Therefore, it is important to clarify the fundamental structure of coal being decomposed effectively by the addition of phenol.

The thermal decomposition of dibenzyl was not affected by the addition of phenol or p-cresol. In contrast, the decomposition of 2,2'-dinaphthyl ether increases remarkably in the presence of phenolic compounds as shown in Table III, and the effect seems to increase with increasing the electron donating property of substituent on the benzene nucleus.

The effect of hydroquinone and p-methoxyphenol is remarkable, but this seems beyond argument at present because considerable parts of them can not be recovered after reaction, suggesting that very complex side reactions are taking place.

TABLE III   EFFECT OF PHENOLS ON THE THERMAL DECOMPOSITION OF 2,2'-DINAPHTHYL ETHER AT 450°C FOR 60 MIN (2,2'-Dinaphthyl ether 11 mmole, Tetralin 225 mmole, Additive 150 mmole)

| | Conversion (%) | |
Additive	Dinaphthyl ether	Phenol
1-Methylnaphthalene	12.6	---
Phenol	17.1	1.5
p-Cresol	21.0	0
2,4,6-Trimethylphenol	26.0	3.5
1-Naphthol	33.7	22.1
p-Phenylphenol	34.1	0
p-Methoxyphenol	49.5	100
Hydroquinone	63.4	57.7

Furthermore, we studied the effect of phenolic compounds
on the thermolysis of phenyl benzyl ether at 320°C, because even
reactive phenols such as p-methoxy phenol are quite stable at
this temperature.

As shown in Table IV, all phenols tested were confirmed to
accelerate the conversion of phenyl benzyl ether. In this
thermolysis, benzyl phenyl ether was decomposed to toluene and
phenol (ca. 60%) and also isomerized to benzyl phenol (ca. 40%).

TABLE IV   EFFECT OF PHENOLS ON THE THERMAL DECOMPOSITION OF
PHENYL BENZYL ETHER AT 320°C FOR 30 MIN
(Phenyl benzyl ether 27 mmole, Tetralin 220 mmole,
Additive 140 mmole)

Additive	Conversion(%) Benzyl phenyl ether	Phenols
1-Methylnaphthalene	31.4	--
p-Cresol	40.2	0
p-Methoxyphenol	44.1	1.1
1-Naphthol	43.0	0

Accelerating Effect due to Phenols on the Rupture of Ether
Linkages. Phenols are weak acids and polar solvent, and so
often observed to enhance the thermal decomposition of covalent
bond, but we could not observe any accelerating effect due to
phenol on the decomposition of dibenzyl.

Therefore, phenols must be participating directly in the
course of scission of ether linkage.

Phenolic compounds may enhance the rate of decomposition
of aromatic ether, because the phenoxy radical may be stabilized
by solvation (18) or hydrogen bonding (19) with phenolic compounds and may result in the subsequent hydrogen transfer
reaction from hydrogen donating solvent or phenols (20).

Previously we have shown that phenolic compounds have a
remarkable positive effect (4) on the coal liquefaction in the
presence of Tetralin, depending strongly on the character of
coal as well as on the concentration of phenols. The effect
of phenols on the decomposition of diaryl ethers will give a
good explanation for the previous results, because aliphatic
ether structures of some young coals will be decomposed rapidly
at relatively low temperatures and so the rate of coal dissolution will not be affected by the addition of phenols, on the
other hand, the polycondensed aromatic ether structures will be
decomposed effectively by the addition of phenols in the course
of coal liquefaction.

The Effect of Mineral Matters on the Decomposition Ethers.
Recently, the effect of mineral matters of coal on the coal
liquefaction has received much attention. It was shown that
small amounts of FeS or pyrite are responsible for the hydro-
genative liquefaction of coal. Therefore, it is interesting
to elucidate the effect of mineral matters of coal on the
decomposition rate and products of aromatic ethers, and so three
diaryl ethers were thermally treated in the presence of coal ash
obtained by low temperature combustion of Illinois No.6 coal at
about 200°C with ozone containing oxygen.

It was found that the addition of coal ash remarkably
accelerates the rate of decomposition of dibenzyl ether and also
drastically changes the distribution of reaction products, that
is, benzyl tetralin becomes the main reaction product instead
of a mixture of toluene and benzaldehyde, as shown in Table V.

This result is quite surprizing, but can be ascribed to the
acidic components of coal ash, since Bell and coworkers (21)
reported that benzyl ether acts as an alkylating reagent in the
presence of Lewis acid such as $ZnCl_2$. However, the acidic effect
due to coal ash may be suppressed considerably by organic base in
the coal liquefaction.

In the cases of phenyl benzyl ether and phenyl 9-phenanthryl
ether, the effect of ash components was not so remarkable.

Further studies on the effect of ash components are under
investigation.

TABLE V EFFECT OF COAL ASH ON THE THERMAL DECOMPOSITION OF
          DIARYL ETHERS (Ethers 4 mmole, Tetralin 40 mmole)

Ethers	Added Ash(g)	Temp. (°C)	Time (min)	Conver- sion(%)	Products(mole% of reacted ether)
Dibenzyl ether	---	400	30	65	$PhCH_3$ 110, PhCHO 77, PhH 13, Benzyltetralin 2.2
	0.015	400	30	100	$PhCH_3$ 6, PhH trace, Benzyltetralin 152
Phenyl benzyl ether	---	400	30	100	$PhCH_3$ 61, PhOH 66, $PhCH_2Ph$ 6, Benzylphenol 27, Benzyltetralin 2
	0.015	400	30	100	$PhCH_3$ 38, PhOH 58, $PhCH_2Ph$ 3, Benzylphenol 39, Benzyltetralin 13
Phenyl 9-phenanthryl ether	---	450	120	65	PhOH 46, Phenanthrene 17
	0.10	450	120	70	PhOH 45, Phenanthrene 18

Conclusions

Diaryl ether must be one of the important structures responsible for the liquefaction of coal among various oxygen-containing organic structures of coal.

The rate of decomposition of polynucleus aromatic ethers increases with increasing the number of nucleus of aryl structure and are enhanced by the addition of phenolic compounds.

The rate of decomposition and the distribution of products of some diaryl ethers can be affected in the presence of coal ash.

Acknowledgement

We acknowledge the financial support from the Sunshine Project Headquarters Agency of Industrial Science Technology.

Literature Cited

1. Curran, G. P., Struck, R. T., Gorin, E., Preprints Amer. Chem. Soc. Div. Petrol. Chem. (1966) C-130-148.
2. Neavel, R. C., Fuel (1976) 55, 237.
3. Farcasiu, M., Mitchell, T. O., Whitehurst, D. D., Preprints Amer. Chem. Soc. Div. Fuel Chem. (1976) 21 (7), 11.
4. Kamiya, Y., Sato, H., Yao, T., Fuel (1978) 57, 681.
5. Fischer, C. H., Eisner, A., Ind. Eng. Chem. (1937) 29, 1371.
6. Takegami, Y., Kajiyama, S., Yokokawa, C., Fuel (1963) 42, 291.
7. Ruberto, R. G., Cronauer, D. C., Jewell, D. M., Seshadri, K. S., Fuel (1977) 56, 17, 25.
8. Ruberto, R. G., Cronauer, D. C., Preprints Amer. Chem. Soc. Div. Petrol. Chem. (1978) 23 (1), 264.
9. Ignasiak, B. S., Gawlak, M., Fuel (1977) 56, 216.
10. Szladow, A. J., Given, P. H., Preprints Amer. Chem. Soc. Div. Fuel Chem. (1978) 23 (4), 161.
11. Washowska, H., Pawlak, W., Fuel (1977) 56, 422.
12. Bacon, R. G. R., Stewart, D. J., J. Chem. Soc. (1965) 4953.
13. Sato, Y., Yamakawa, T., Onishi, R., Kameyama, H., Amano, A., J. Japan Petrol. Inst. (1978) 21, 110.
14. Benjamin, B. M., Raaen, V. F., Maupin, P. H., Brown, L. L., Collins, C. J., Fuel (1978) 57, 269.
15. Brower, K. R., Fuel (1977) 56, 245.
16. Pott, A., Broche, H., Fuel in Science and Practice (1934) 13 (4), 125.
17. Orchin, M., Storch, H. H., Ind. Eng. Chem. (1948) 40, 1385.
18. Huyser R. S., Van Scoy, R. M., J. Org. Chem. (1968) 33, 3524.
19. Howard, J. A., Furimsky, E., Can. J. Chem. (1973) 51, 3788.
20. Mahoney, L. R., Da Rooge, M. A., J. Amer. Chem. Soc. (1967) 89, 5619.
21. Mobley, D. P., Salim, S., Tanner, K. I., Taylow, N. D., Bell. A. T., Preprints Amer. Chem. Soc. Div. Fuel Chem. (1978) 23 (4), 138.

RECEIVED March 28, 1980.

Possible Hydride Transfer in Coal Conversion Processes

DAVID S. ROSS and JAMES E. BLESSING

SRI International, 333 Ravenswood Avenue, Menlo Park, CA 94025

The conversion of coal to liquid fuels is usually carried out in the presence of an H-donor solvent (H-don) such as Tetralin. The chemical route commonly suggested for the process is

$$coal\text{-}coal \rightleftharpoons 2 \ coal\bullet \qquad (1)$$

$$coal\bullet \xrightarrow{\text{H-don}} coalH \qquad (2)$$

in which there is initial thermal homolysis of sufficiently weak bonds in the coal structure to yield radical sites. These reactive sites are then "capped" by transfer of hydrogen atoms from the donor solvent. We will discuss here the chemistry of this process, including detailed consideration of the thermochemistry of Steps 1 and 2 above. We will then present some of our recent data, which suggest that there may be an ionic component in the process.

Background

Step 1 above requires that there be bonds in the coal that are weak enough to break in appropriate numbers at conversion temperatures and times. Table I displays some kinetic data for the cleavage of benzylic bonds in a series of increasingly aromatic compounds. In accord with expectation, an extension of the aromatic system increases the ease with which the benzylic bond is broken. The phenanthrene system appears to be no more easily cleaved than the naphthalene system; however, ethyl anthracene is clearly destabilized significantly more than the other compounds in the table. The large decrease in bond-dissociation energy for the anthracene system is reflected in the increase by three to four orders of magnitude in the rate of scission at conversion temperatures, as shown in the table.

Also pertinent to Step 1 is the material in Table II, which includes bond dissociation energies and kinetic data at conversion temperatures for a series of C-C bonds. For the purposes of this discussion it can be assumed that substitution of -O- for -CH$_2$-

0–8412–0587–6/80/47–139–301$05.00/0

Table I

THERMAL CLEAVAGE OF BENZYLIC BONDS IN AROMATIC SYSTEMS

Structure	BDE[a] (kcal/mol)	Relative Rate	
		400°C (572°F)	500°C (932°F)
CH_2-CH_3 (benzene)	72 (obs.)	1	1
CH_2-CH_3 (naphthalene)	68 (obs.)	20	12
CH_2-CH_3 (phenanthrene)	68 (est.)	20	12
CH_2-CH_3 (anthracene)	60 (est.)	9×10^3	3×10^3

[a] Bond dissociation energies. See S. Stein and D. Golden, J. Org. Chem., 42, 839 (1977). The two observed values correspond to estimated values, and are from the unpublished work of D. Golden and D. McMillen, SRI.

Table II

THERMOCHEMICAL AND KINETIC FACTORS FOR SOME BONDS IN COAL

Structure	BDE[a] (kcal/mol)	$t_{1/2}$[b] 400°C (572°F)	500°C (932°F)
	> 120	∞	∞
	116	2.6×10^{15} yr	3.7×10^{10} yr
	82	2.7×10^4 yr	10.2 yr
	81	1.3×10^4 yr	5.3 yr
	69	629 days	19.7 hr
	57	2.0 hr	30 sec

[a]Bond dissociation energies. See S. W. Benson, Thermochemical Kinetics, John Wiley and Sons, New York (1968).

[b]Calculated from the expressions log k (sec^{-1}) = 14.4 - BDE/4.6T, $t_{1/2}$ = 0.7/k.

does not change the thermochemistry. Thus, for example, the Ph–OPh bond and the Ph–CH$_2$Ph bond are similar in strength. Again not surprisingly, double benzylic resonance as present upon the scission of the central bond in bibenzyl results in a significant destabilization, and that compound, along with benzyl phenyl ether, PhOCH$_2$Ph, are the only compounds with bonds easily broken under conversion conditions. If the increase in rate by a full three to four orders of magnitude (shown in Table I) for anthracene systems is applied, we see that perhaps the 1,3-diarylpropane system may also be sufficiently unstable for conversion at 400°–500°C. Thus it would seem that for coal conversion via Steps 1 and 2, at least at 400°C, Step 1 can be sufficiently rapid for some structural features. We will discuss below some conversion data at 335°C, however, which suggest that thermally-promoted bond-scission is not fully consistent with experimental observation.

Next, a consideration of the kinetics for Step 2 raises some questions. The transfer of hydrogen in similar reactions has been well studied, and Table III presents data for the relative rates of transfer of hydrogen from a number of hydrocarbons to the free radical Cl$_3$C• at 350°C. The donor hydrocarbons are listed in order of increasing ease of H-transfer to the free radical. Tetralin is near the middle of the list. The most reactive donor in the table, 1,4-dihydronaphthalene, is about four times as active as Tetralin.

The table also shows the results of experiments with the donors and coal in phenanthrene as solvent. Consistent with the transfer of hydrogen in a radical process, those donors less reactive toward Cl$_3$C• than Tetralin are also less effective than Tetralin in conversion of coal to a phenanthrene-soluble product. However, in contrast to the chemistry of Step 2 we see that those donors that are more reactive toward Cl$_3$C• than Tetralin are also less effective in their action with coal. Thus this simple conversion scheme is suspect.

Current Results

Alcoholic KOH. We have reported on the use of isopropyl alcohol as an H-donor solvent in coal conversion, and specifically on the effects of the addition of strong bases such as KOH to the system (1a). We found that i-PrOH brought about a conversion of Illinois No. 6 coal very similar to the conversion level obtained by Tetralin under the same conditions. These results are listed in Table IV in text, along with the results of more recent experiments using methanol as the solvent and adding KOH to the system (1b).

Isopropanol is of course a well known reducing agent under basic conditions, reducing carbonyl compounds via hydride transfer, and becoming oxidized to acetone in the process (2). The table shows that the addition of KOH to the system significantly increases the effectiveness of the coal conversion reaction, and it would seem that such a system would have an advantage over one based on Tetralin, where significant catalysis of hydrogen transfer

Table III

REACTIVITY OF VARIOUS H-DONOR SOLVENTS WITH CCl_3 AND COAL

Cosolvent	Relative Reactivity Toward CCl_3 [a] (350°C)	Relative % Dissolution[b] (2 hr/350°C)
None (Phenanthrene)	--	[33] [c]
Toluene	1.0	1
2-Methylnaphthalene	2.1	0
1-Methylnaphthalene	2.6	2
Diphenylmethane	6.0	1
Cumene	8.7	0
Fluorene	(20)	0
Tetralin	41.0	27
9-Methylanthracene	56.0	0
1,2-Dihydronaphthalene	(65)	6
9,10-Dihydroanthracene	(102)	24
9,10-Dihydrophenanthrene	(102)	22
1,4-Dihydronaphthalene	(160)	14

[a] Data from Hendry, D., Mill, T., Piszkiewicz, Howard, J., and Eigenman, H., J. Phys. Chem. Ref. Data, 3, 937 (1974). The values in parenthesis are estimated from other radical data available in the paper. [b] All reactions carried out in phenanthrene as solvent with 4.0 parts solvent per part coal, and 10 wt % cosolvent. Unpublished data of D. Hendry and G. Hum, SRI. [c] Value for phenanthrene alone.

has not been directly demonstrated. We found in our experiments
with i-PrOH/KOH at 335°C that coal was converted to a product about
60% soluble in i-PrOH, that fraction having a number-average molecular weight of about 460.

In some model compound studies with the i-PrOH/KOH system we
found that anthracene was converted to 9,10-dihydroanthracene in
64% yield. Benzyl phenyl ether was also studied and was converted
to a polymeric material under the reaction conditions. There were
no traces of phenol nor toluene, the expected reduction products.

We found subsequently that MeOH/KOH media at 400°C were very
effective reducing systems, as Table IV shows (1b). The methanol
work yielded products with significant reductions in organic sulfur
levels and moderate reductions in nitrogen levels. We suggest the
mechanism of reduction is ionic in nature, involving hydride transfer. Thus

$$H-\overset{|}{\underset{|}{C}}-OH + OH^- \longrightarrow H_2O + H-\overset{|}{\underset{|}{C}}-O$$

$$H-\overset{|}{\underset{|}{C}}-O^- + coal \longrightarrow \overset{|}{\underset{|}{C}}=O + coalH^-$$

$$coalH^- + H-\overset{|}{\underset{|}{C}}-OH \longrightarrow H-\overset{|}{\underset{|}{C}}-O^- + coalH_2$$

where coalH$^-$ and coalH$_2$ are an anionic intermediate and reduced
coal, respectively. The net reaction is

$$H-\overset{|}{\underset{|}{C}}-OH + coal \longrightarrow \overset{|}{\underset{|}{C}}=O + coalH_2$$

and in fact, in the i-PrOH work, we isolated acetone in quantities
consistent with the quantities of hydrogen transferred to the coal.

The final coal product in the MeOH/KOH experiments was 20%-25%
soluble in the methanol. When the methanol was removed, the resultant product was a room temperature liquid with the properties
described in Table V. Apparently the polymethylphenol fraction is
formed by the cleavage of phenolic ethers and subsequent methylation by the CO that is present in the reaction mixture as a result
of methanol decomposition. The methylation reaction has been
observed before for similar systems (3).

The methanol-insoluble product was also upgraded relative to
the starting coal. Its H/C ratio was 0.86, its sulfur and nitrogen
levels were 0.8% and 1.2%, respectively, and it has fully pyridine
soluble.

Model compound studies were also carried out in MeOH/KOH, and
the results are shown in Table VI. Phenanthrene and biphenyl were
quantitatively recovered unchanged by the reactions, and bibenzyl
was recovered in 95% yield, with small amounts of toluene observed.
Anthracene and diphenyl ether, on the other hand, were converted
respectively to 9,10-dihydroanthracene and a mixture of polymethylphenols similar to that observed in the work with coal. The cleavage of diphenyl ether via hydrogenolysis should yield both benzene
and phenol as products; we saw no benzene in our study, and our

Table IV

ALCOHOLS AS H-DONOR SOLVENTS IN COAL CONVERSION

System	T(°C)/min	Pyr. Sol (%)	H/C	%S org	%N
Untreated coal	--	13	0.73	2.1	1.7
Tetralin	335/90	48	0.81	1.8	1.6
i-PrOH	335/90	50	0.81	1.8	1.6
i-PrOH/KOH	335/90	96	0.88	1.6[a]	1.6
MeOH/KOH	400/90	99[+]	0.96	0.8	1.2

[a]Reduced to 0.5% when product treated further.

Table V

PROPERTIES OF THE METHANOL-SOLUBLE COAL PRODUCT

H/C   1.37 ⎫
                 ⎪   ~40% polymethylphenols and anisoles
%S org   0.5   ⎪ distilled   ~40% mixture of butyrolactone and
                 ⎬ ─────────→   heavier materials yielding
%N   0.4   ⎪ <0.1 torr $CH_3CH_2CH_2C=O^+$ and
                 ⎪ $HOCH_2CH_2CH_2C=O^+$ major
H_{al}/H_{ar}[a] 5.1 ⎭   fragments by mass spectroscopy

                           ~20% unidentified

[a]By nmr

Table VI

MODEL COMPOUNDS IN MeOH/KOH AT 400°C/30 MINUTES

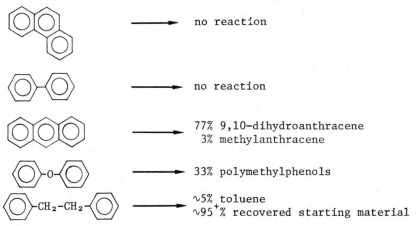

no reaction

no reaction

77% 9,10-dihydroanthracene
3% methylanthracene

33% polymethylphenols

~5% toluene
~95[+]% recovered starting material

observation of the polymethylphenols and anisoles thus indicates
that nucleophilic ether cleavage is taking place. In other words,
it appears that phenoxide is displaced by methoxide, thus,

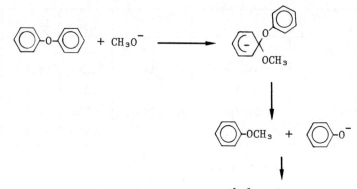

methylated products

so that a favorable feature of the MeOH/KOH system, in addition to
its reducing power, is its facility in cleaving otherwise inert
ethers.

CO Chemistry. The alcohol/base chemistry observed here led
logically to a system including CO/H₂0/KOH, and accordingly, a
series of experiments was performed at 400°C. The COSTEAM Process
is similar in nature, but without the purposeful addition of base.
Also, the process is applied primarily to lignite, though the CO-
STEAM chemistry has been applied, less successfully, to bituminous
coal also. The results we obtained with the basic system, along
with the pertinent citations to earlier work by others, were
recently presented (4)
     To summarize our effort, we found that CO/H₂0/KOH systems con-
verted Illinois No. 6 coal to a material which was fully pyridine
soluble, 51% benzene soluble, and 18% hexane soluble. As with the
basic alcoholic systems, there were significant reductions in organic sulfur levels, and moderate reductions in nitrogen levels.
The chemistry here is similar to that for the basic alcoholic
systems, but with formate (HCO₂⁻) as the hydride-donor, and thus
the reducing agent. In control runs with H₂0/KOH or CO/H₂0, little
or no conversion was observed.

Model Component Studies. Model compound work with this system
showed that anthracene was reduced to its 9,10-dihydro derivative
(35% yield). Bibenzyl, on the other hand, was recovered unchanged,
with only a trace of toluene observed.
     The model compound work for the three basic systems is summa-
rized in Figure 1. A finding of no significant reduction in a
system is designated by an x'ed arrow. Our criterion of successful
reduction requires that significant quantities of the starting

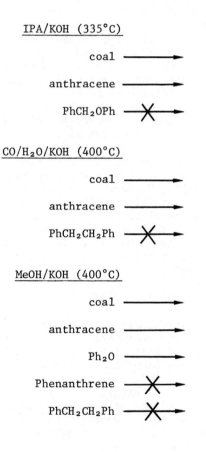

IPA/KOH (335°C)

coal ⟶

anthracene ⟶

PhCH₂OPh ⟶✗⟶

CO/H₂O/KOH (400°C)

coal ⟶

anthracene ⟶

PhCH₂CH₂Ph ⟶✗⟶

MeOH/KOH (400°C)

coal ⟶

anthracene ⟶

Ph₂O ⟶

Phenanthrene ⟶✗⟶

PhCH₂CH₂Ph ⟶✗⟶

Figure 1. Summary of results for coal and selected model compounds in some strongly basic conversion systems. Arrows indicate that significant conversion to reduced products was observed. The X'ed arrows indicate that no reduction was observed.

material be converted to reduced products. With benzyl phenyl
ether, for example, while little starting ether was recovered, most
of it being converted to an intractable polymer, no phenol nor
toluene was found. Thus we conclude the system was not effective
in reduction of the C-O bond. Similar statements apply regarding
the C-C bond in bibenzyl for the MeOH/KOH and CO/H$_2$O/KOH systems.

Discussion

A significant conclusion from these data is that coal is converted under conditions where the common model compounds benzyl
phenyl ether and bibenzyl are not reduced. In explanation, it
might be suggested that there are two conversion mechanisms. One
would be the commonly considered scheme (Steps 1 and 2), taking
place in Tetralin-like media, involving free radical chemistry, and
reducing both coal and such model compounds as bibenzyl and phenyl
ether through a thermally-promoted initial bond-scission. A second
mechanism would be operative in strongly basic media, involve hydride transfer, and would perhaps include the conversion of coal
via chemistry related in some way to the reduction of anthracene
by these systems.

However, as pointed out above, the commonly proposed free
radical mechanism is not entirely consistent with the observed
behavior of H-donor solvents and coal. Further, a thermally promoted C-C or C-O bond-scission is inconsistent with our observations in the i-PrOH work at 335°C. As also mentioned, a major
fraction of the coal was converted in this system to a product with
a number-average molecular weight of less than 500. If we consider
that the rate constant for the unimolecular scission of the central
bond in bibenzyl is expressed (5) as

$$\log k \ (\text{sec}^{-1}) = 14.4 - 57/2.303RT$$

then, while the half-life for the reaction at 400°C is about 2
hours, and thus perhaps appropriate for considerations of conversion at those conditions, at 335°C the half-life is about 160 hours,
and clearly the reaction cannot play a significant role in the conversion of coal at that temperature.

Additionally, it has been noted that Tetralin operates via
hydride transfer, at least in its reduction of quinones. Thus it
has been shown that Tetralin readily donates hydrogen to electron-
poor systems, such as quinones at 50°-160°C. The reaction is
accelerated by electron-withdrawing substituents on the H-acceptor
and polar solvents, and is unaffected by free radical initiators
(6). These observations are consistent with hydride transfer, as
is the more recent finding of a tritium isotope effect for the reaction (7).

We propose, therefore, that the operative mechanisms of coal
conversion in both Tetralin-like media and our strongly basic
systems may be the same, involving hydride donation by the H-donor

solvent, followed by proton transfer. Consistent with this surmise
are the results from two experiments carried out under the same
conditions, utilizing Tetralin on the one hand, and $CO/H_2O/KOH$ on
the other. The results, presented in Table VII, show that the

Table VII

COMPARISON OF $CO/H_2O/KOH$-TREATED AND TETRALIN-TREATED
COAL PRODUCTS[a]

Reaction Conditions	Benzene Solubility[b] %	Molar H/C		%N		H_{al}/H_{ar} [e]
		BS[c]	BI[d]	BS[c]	BI[d]	
$CO/H_2O/KOH$[f]	48	1.01	0.65	1.1	1.8	∿ 3.0
Tetralin[g]	37	1.01	0.72	0.6	1.9	∿ 2.7

[a]Reactions run at 400°C for 20 min in a stirred, 300 ml Hastelloy
"C" reactor. 10.0 g of dried, –60 mesh Illinois No. 6 coal (PSOC-
26) used in each case. [b]Solubility of entire product in 50 ml of
benzene. [c]Benzene soluble portion of the product. [d]Benzene insoluble portion of the product. [e]Ratio of ^{1}H-NMR areas: H_{ali} ≡
$0<\delta<5$; H_{arom} ≡ $5<\delta<10$. Contributions from benzene and Tetralin
were substrated before calculation. [f]36 g of H_2O, 10 g of KOH,
and 700 psig of CO used in a 300 cc reactor. [g]60 g of Tetralin.

products from the two reactions have similar diagnostic characteristics, including benzene solubilities, H/C ratios, and ratios of
H_{al}/H_{ar}. Additionally, HPLC profiles for the two products are very
similar.

Since the suggested conversion process does not include a thermally promoted bond-scission step, the question arises of how the
addition of hydrogen results in the bond breaking necessary for
significant reduction in molecular weight. We have already noted
that the nucleophilic action of the basic methanol system was
sufficient to cleave diphenyl ether, and a similar route is available in the basic i-PrOH and CO/H_2O systems. On the other hand,
we showed in control experiments that strongly basic conditions
alone were not sufficient for significant conversion of coal.

On the basis of the data at hand, we are currently considering
two possible modes of molecular weight reduction. The first involves the generation of thermally weak bonds by the initial addition of hydrogen. We suggest that the addition of hydrogen to the
structures below may be a key to the cleavage of critical bonds in
coal.

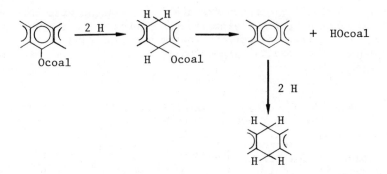

It can be shown thermochemically that the addition of hydrogen across structures like those above is favored under conversion conditions (1a). In turn, it can be suggested that the dihydro-ether intermediate is rapidly thermalized in the succeeding step, yielding both an oxygen-containing fragment and a rearomatized fragment that is rapidly reduced to a hydroaromatic product. The thermolysis of the intermediate is expected to be rapid at conversion temperatures, in accord with Brower's observation that, in Tetralin, anthraquinone is converted all the way to anthracene (8). Moreover, it is recognized that 9-hydroxy-9,10-dihydroanthracenes readily eliminate water at ambient conditions, yielding the aromatized product (9).

The second potential conversion mode takes into consideration the recent observation that hydrogen acceptors such as benzophenone oxidatively couple phenols under conversion conditions (10). It can thus be suggested that the major role of a reducing component in a coal conversion system is the reduction of quinones and perhaps other oxidants present in the coal, rather than direct reduction of the coal. In the absence of an H-donor, then, oxidative crosslinking takes place within the coal upon heating, yielding a product even less soluble in solvents such as pyridine than was the starting coal. On the other hand, in the presence of a reducing agent, either conventional H-donors such as Tetralin or our hydride-donating systems, the quinones and other oxidants should be reduced to unreactive material, and the coal might then proceed to liquefy by means of a thermal process involving no addition of hydrogen. We cannot at this time propose a route for purely thermal liquefaction (reverse Diels Alder reactions might be suggested for purposes of example), and the concept currently remains a working hypothesis.

Note. In a recent paper, Miller and Stein have provided values for both C-C and C-O bonds for a variety of coal model compounds, including bibenzyl and benzyl phenyl ether (11). Their rate constant for bibenzyl provides half-life values at 335°C and 400°C even larger than those discussed here, and it would seem on the basis of their data that at those low temperatures C-C scission in bibenzyl itself is too slow for thermal scission to be significant.

On the other hand, they find that the ether and substituted bi-
benzyls are thermalized faster than bibenzyl itself, and thus thermal scission of such bonds could play a role in a free radical
route.

Acknowledgement

We acknowledge the support of this work by the U.S. Department
of Energy under Contract No. EF-76-01-2202.

Literature Cited

1. (a) Ross, D. S.; Blessing, J. E., Fuel, 1979, 58, 433-437;
    (b) Ibid., pp. 438-442.

2. Wilds, A., "Organic Reactions"; Vol. II, 180, 1947.

3. Stenberg, V., private communication.

4. Ross, D. S., and Blessing, J. E., Fuel, 1978, 57, 379.

5. Benson, S. W., and O'Neal, H., "Kinetic Data on Gas Phase
    Unimolecular Reactions"; National Bureau of Standards, NSRDS-
    NBS 21, 1970.

6. Linstead et al., J. Chem. Soc., (1954) 3548, 3564.

7. Van der Jagt, P.; de Haan, H.; and van Zanten, B.; Tetr. 1971,
    27, 3207.

8. Brower, K., Fuel, 1977, 56, 245.

9. Fieser, L.; and Hershberg, R., J. Am. Chem. Soc., 1939, 61,
    1272.

10. Raaen, V.; and Roark, W., Fuel, 1978, 57, 650.

11. Miller, R.; and Stein, S., Fuel Division Preprints for the
    178th National Meeting of the American Chemical Society, 1979,
    Washington, D.C., Sept. 10-14, p. 271.

RECEIVED March 28, 1980.

Hydrogenation Mechanism of Coals by Structural Analysis of Reaction Products

Y. MAEKAWA, Y. NAKATA, S. UEDA, T. YOSHIDA, and Y. YOSHIDA

Government Industrial Development Laboratory, Hokkaido,
41–2 Higashi–Tsukisamu Toyohiraku, Sapporo, 061–01, Japan

It is generally conceded that in the hydrogenation reaction of coal, the following diverse chemical reactions compete in parallel (among and against each other) as the reaction proceeds: namely, thermal decomposition, stabilization of active fragments by hydrogen, cleavage of linkages between structural units, dealkylation, dehetro atom, hydrogenation of aromatic rings, ring opening, etc. It is also accepted that the various features of these reactions are strongly influenced and vary with the type of raw coal, their individual chemical structures, properties of the catalyst employed, type of reducing agents, conditions of reaction; temperature and pressure and the degree of reaction progress.

The resulting reaction products are a complex mixture of compounds and because of the fact that the structural analysis thereof is extremely complicated, it follows that elucidation and clarification of the reaction mechanisms involved are extremely difficult and the results are far from satisfactory.

Thus we have conducted work on the structural parameters of coal hydrogenation products using the method of Brown-Ladner(1), and from the results obtained we have developed correlations of the reaction. Based on the above, the outline of the reaction mechanisms have been previously discussed and our results have been reported (2,3).

In the present paper samples using several kinds of coal and various reducing agents such as H_2, H_2 + CH_4, D_2, D_2 + tetralin, CO + H_2O, we have carried out hydrogenation reactions. We have studied the distribution and structural parameters of the reaction products and we have further discussed the reaction mechanisms involved.

Experimental

The analysis of the sample coals used in the experiments are shown in Table 1.

Hydrogenation and reduction were conducted using a batch type

Table 1   Analytical data on the coals studied

Sample	Proximate analysis %				Ultimate analysis %				
	M.	Ash	V.M.	F.C.	C	H	N	S	O
Soya Koishi	15.5	15.7	33.6	35.3	73.0	6.6	1.5	0.04	20.0
Bayswater Vi[1]	3.4	1.6	32.9	62.1	83.0	5.3	2.0	0.5	9.2
Bayswater In[2]	4.5	16.2	20.8	58.5	85.0	4.1	1.9	0.3	8.7
Yubari	1.1	6.8	43.6	48.5	85.2	6.2	1.6	0.1	6.9
Shin Yubari	1.2	7.4	34.7	56.7	87.4	6.5	1.8	0.04	4.7

[1] Australian, Vitrinite 99% concentrate
[2] Australian Inertinite 95% concentrate

autoclave with an inner volume of 500 ml. The reaction gas after
completion of the reaction was analysed by gas chromatography.
Further, the produced water was quantitated. The remaining
products in its entirety were quantitated and fractionated by
extraction using n-hexane, benzene, and pyridine. The fractionation methods are shown in Fig. 1. Regarding these, when fractionation is accomplished by benzene extraction, the conversion was
calculated from organic benzene insolubles (O.B.I.) and when
fractionation is carried out by pyridine the same was calculated
from organic pyridine insolubles (O.P.I.)

    With regards to hydrogenation, reducing agents such as H_2,
$H_2 + CH_4$, D_2, D_2 + Tetralin, $CO + H_2O$ were selected and reduction
was conducted by varying the reaction time. Each isolated
fraction was subjected to ultimate analysis, H-NMR, C-13 NMR,
molecular weight measurement and the structural parameters were
calculated. The results of the study of these structural parameters in the course of the reactions were evaluated and the
reaction mechanisms thereof are discussed below.

Results and discussion

Product distribution   For many years high pressure hydrogenation
reaction has been dealt with as a consecutive reaction with
asphaltene as the intermediate (4,5,6). Further it has been
pointed out that Py-1, O_2 likewise shows the behavior of intermediates. (See Figure 1) (3).

    In Fig. 2 the following is depicted; the hydrogenation pro-
ducts of Yubari coal at a reaction temperature of 400°C were
fractionated by the procedure shown in Fig. 1, and the changes
of these fraction yields by reaction time are shown. Inasmuch as
the presence of Py-2 is scarce in coal, the observed results
were obtained in the decreasing period, thus it may readily be
surmised that Py-2 must be produced in the initial stages of
the reaction. Here we consider oil-1 as the final product, and
it may be noted that all fractions other than oil-1, show an initial

e of the reaction
intermediate

ogenation of
ian Bayswater coal
g. 5 shows the
Soya-Koishi coal
milar tendency to
ndicates that the
ate products in a

ation and reduc-
ider range of
he hydrogenation
e reaction which

e distribution of
that these have
also known that
0). Therefore,
drogenation of
oal gradually
also underway.
ressed in a

ane Soxhlet
ction

ene Soxhlet
action

enzene
-O.B.I.)

idine extraction
ss filter G_4)

Residue

Pyridine Soxhlet
extraction

Organic pyridine
insolubles
(O.P.I.)

$$\cdots \rightarrow Fr\ N_1$$
$$\cdots \rightarrow Fr\ N_2$$
$$\cdots \rightarrow Fr\ N_n$$
$$\cdots \rightarrow Fr\ N_{n+1}$$

$$Fr\ N_{n+n}$$

ed and leads to
A_1, and at the
y produced.
tituents have
ferent compo-
reaction may
The method
owering direc-
, and it may be

tion of coal hydrogenation product

increase and a subsequent decrease in the cours
and it is clear that they show the behavior of
products in a consecutive reaction.

The distribution of products from the hydr
vitrinite and inertinite separated from Austral
is shown in Fig. 3 and Fig. 4 respectively. Fi
distribution of products from the reduction of
by CO + H_2O. The product distribution shows a s
that of Yubari coal shown in Fig. 2. Thus, it
Py-2, Py-1, A, O_2 can be considered as intermedi
consecutive coal hydrogenation reaction scheme.

We used 3 solvents to separate the hydroge
tion products into fractions. If we can use a w
solvents with different solubility properties, t
reaction scheme can be expressed as a consecutiv
has more steps.

Based on our pervious work (7, 8) regarding th
molecular weight of these fractions, it is known
a wide distribution of molecular weight. It is
raw coal itself has a structural distribution (9,
it may be considered that in the high pressure h
coal, the reactivity of the remaining unreacted
changes while consecutive degradation of coal is

The consideration mentioned above may be exp
schematic diagram as follows.

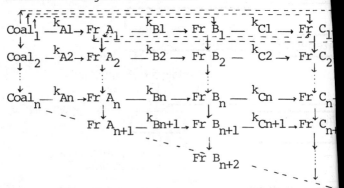

In other words, $coal_1$ (raw coal) is reac
a change to a one step lower molecular fraction Fr
same time as a side reaction Fr B_1 Fr N_1 is direct
$Coal_2$, (the remaining coal from which the above con
been removed from $coal_1$), would naturally have a di
sition and reactivity from that of $coal_1$. A similar
be expected when Fr A_1 undergoes a change to Fr B_1
of dealing with the above indicates the molecular l
tion of the reaction of hydrogenation and reduction

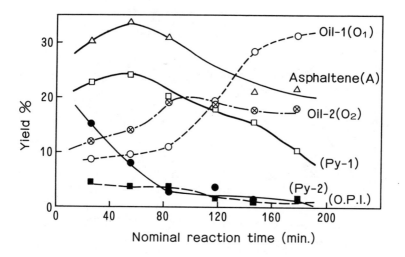

Figure 2. Distribution of products from Yubari coal hydrogenation at 400°C

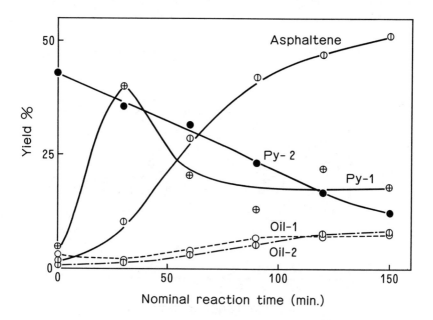

Figure 3. Distribution of products from Bayswater vitrinite hydrogenation at 400°C

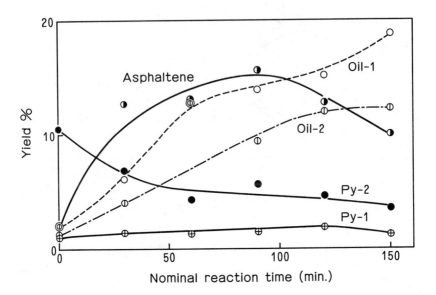

Figure 4.   *Distribution of products from Bayswater inertinite hydrogenation at 450°C*

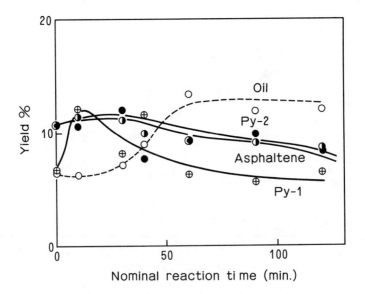

Figure 5.   *Distribution of products from Soya–Koishi coal liquefaction with CO + H₂O at 400°C*

surmized that the chemical reaction of each step would be highly complex.

Based on the above considerations, the mode of kinetic study carried over a wide range, is solely a means to express how the fraction, isolated by a certain method, changes quantitatibely in the reaction course. In other words these experimental results are expressed as an equation for convenience sake. And it may be considered as a practical means in applicable form.

Now, since each respective fraction shows a structural change along with the reaction course, it may be considered at the same time that it contains products from a higher molecular fraction. In other words, for instance Fr A$_2$ is produced from coal 2 which changes by hydrogenolysis and at the same time, the residue resulting from the reaction from Fr A$_1$ to Fr B$_1$ also appears in this fraction.

This should be taken into consideration when the reaction mechanism is to be discussed. In Fig. 2, it may be appropriate to consider that the structural changes of Py-1 and Py-2 in the course of the reaction are due to the reaction itself.

Reaction mechanism of coal hydrogenation by H$_2$

Hydrogenolysis was conducted using Yubari coal at a reaction temperature of 400°C. The changes of ultimate composition and structural parameters of the fractions of the hydrogenation products in the reaction course are shown in Fig. 6,7. The oxygen clearly decreases as the consecutive reaction proceeds while the hydrogen increases. Whereas the nitrogen level in the oil 2, asphaltenes, Py-1 and Py-2 fraction is greater than that in the original coal, it appears that the nitrogen containing structure is refractory against hydrogenation. The structural parameters calculated by the Brown-Ladner's H-NMR method are shown in Fig. 7. Hydrogenation reaction proceeds from Py-1 to O$_2$, therefore, fa is found to decrease as the reaction progresses while both Hau/Ca and σ increase. This means that saturation of the aromatic ring was occuring.

The molecular weights of the Py-1, A, and O$_2$ fractions were 2000, 800 and 350 respectively whereas the weights of the structural unit, as calculated from the structural parameters were about 300, 300 and 280 respectively. This indicates that cleavage of the linkages between structural units has occurred. The decrease in Hau/Ca of Py-1 in the latter part of the reaction, even when considered together with the behavior of other parameters, cannot be considered to be the result of growth of the condensed aromatic ring of Py-1. It is assumed that because of the Brown-Ladner assumption is lacking the diphenyl type linkage, Py-1 has it. Generally, the Hau/Ca calculated Brown-Ladner's method is lower than that from C-13 NMR in the case of diphenyl type linkage containing substances (11). Therefore, it is doubtful that the Hau/Ca from H-NMR directly reflects the degree of aromatic ring

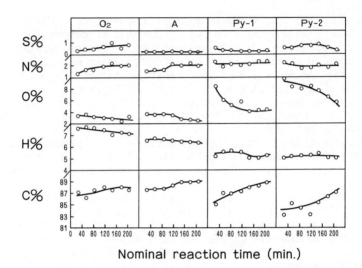

Figure 6.   Ultimate composition of Yubari coal hydrogenolysis product

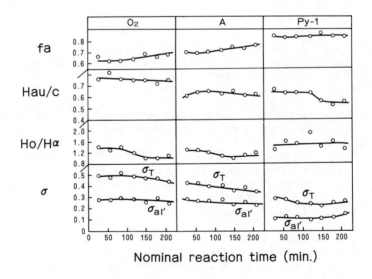

Figure 7.   Structural parameters of Yubari coal hydrogenolysis products

condensation but the tendency can be accepted.

Although it can be stated that (by the decrease of the molecular weight), the Py-l is transferred to asphaltene, when the structural parameters of Py-l and asphaltene are compared, it may be conceded that at this molecular weight lowering step, deoxygen reaction, depolymerization ring saturation (side chain substitution - al') and likewise an increase in side chain substitution over O_2 etc are taking place. Similar results may be observed in the hydrogenation of Bayswater vitrinite concentrate, and inertinite concentrate.

Reaction mechanism of coal hydrogenation by $H_2 + CH_4$

It was noted that CH_4 was produced as the largest gas product in the hydrogenation of coal and is found coexisting with reducing hydrogen. Thus the influence therof on the reaction mechanism was investigated (12). Methane was added at a pressure of 25 or 50 kg/cm^2 to the initial hydrogen pressure of 75 kg/cm^2 and the hydrogenation reaction rate constant was measured. The results are as shown in Table 2.

Table 2   Reaction rate constant for Shin-Yubari coal hydrogenolysis under different reducing gas compositions

Reaction Temp.	Reducing gas composition (kg/cm^2)	Reaction rate constant min.$^{-1}$ $K_1 + K_3$	K_1	K_3
	H_2 75	0.0115	0.0037	0.0078
400°C	H_2 75 + CH_4 25	0.0133	0.0050	0.0083
	H_2 75 + CH_4 50	0.0111	0.0033	0.0078
	H_2 75 + Ar 25	0.0109	0.0024	0.0085
	H_2 50 + CH_4 50	0.0178		
450°C	H_2 75 + CH_4 25	0.0230		
	H_2 100	0.0258		

It may be seen that the reaction rate constant mainly depends on the hydrogen partial pressure. However, as compared with the hydrogen only where methane is added at a pressure of 25 kg/cm^2 is a slightly higher value seen.   But, in the case of 450°C likewise it may be considered that it solely depends on hydrogen partial pressure. The results of an investigation on the change of structural parameters of the reaction products by reaction time are shown in Fig. 8. Thus, it may be considered that the increase in hydrogen pressure mainly enhances the saturation of aromatic rings (as a result an increase in α-hydrogen is seen).

Reduction mechanism of coal by D_2 and D_2 + Tetralin

In order to clarify the attack site of the hydrogen used for reduction, D_2 was used as the reducing agent and deutrization of coal was conducted and an investigation of the D distribution of reaction products was carried out (13). For the measuring distribution of D, D-NMR was used, and determination of D divided in 3 categories; namely, aromatic D(Da), D bonded carbon α from aromatic ring (Dα), and D bonded carbon further β from aromatic carbon. The results are as shown in Table 3.

Da, Dα, Do used together with H gave an approximately similar hydrogen type distribution to that when H_2 only was used. However, when compared with the distribution of hydrogen remaining in the reaction products, it was shown that a marked maldistribution of deuterium at the α position existed. It was also noted that when Tetralin is used as the vehicle, this tendency became more pronounced.

Table 3   H and D distribution of hydrogenation and deutration product of Shin-Yubari coal at 400°C under 50 kg/cm^2 of initial pressure

Product	Gas	Rt(min.)	H and D distribution %					
			Ha	Hα	Ho	Da	Dα	Do
Oil	D_2	60	17.5	19.5	63.0	19.4	49.5	31.1
	D_2	120	18.7	21.2	60.1	19.9	50.9	29.2
	H_2	60	21.3	29.6	49.1			
	H_2	120	19.8	31.2	48.9			
	D_{2*}	60	28.8	29.8	41.4	20.1	54.5	25.4
	H_{2*}	60	30.5	35.3	34.2			
Asphaltene	D_2	60	30.3	29.9	39.8	12.9	61.7	25.4
	D_2	120	26.5	29.9	43.5	12.5	55.1	32.4
	H_2	60	27.4	36.4	36.2			
	H_2	120	30.7	34.5	34.8			
	D_{2*}	60	26.1	29.8	44.2	–	68.3	31.7
	H_{2*}	60	29.5	34.2	36.3			

* with Tetralin

In addition to the minute amounts of hydrogen in the produced gas, corresponding to the maximum 34% of hydrogen in coal is present as H-D, and it is known that regarding D in the reaction products, not only D from the reaction but also D arising from the H-D exchange reaction are present. While there is a strong selectivity of H-D exchange reaction (14), the ratio of Da / Do is comparatively high in the products of the initial stage and further even with the increase in reaction time, since the Da / Dα / Do

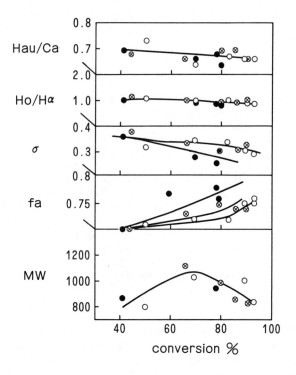

Figure 8. Distribution of structural parameters of asphaltene from Shin–Yubari coal hydrogenation at 450°C: (○), H₂:100kg/cm²; (⊗), H₂:75 + CH₄:25; (●), H₂:50 + CH₄:50

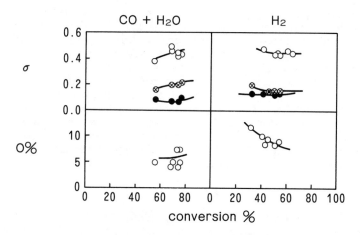

Figure 9. Structural parameters of asphaltene from Soya–Koishi coal by CO + H₂O and H₂ reduction at 400°C

ratios do not approach the Ha/H/Ho ratios, it may be considered that a larger portion of D reacts to carbon from the aromatic rings.

Reduction by CO + H$_2$O

In Fig. 9 is shown a comparison of changes of the structural parameters of the reaction products in the reaction course where reduction of Soya coal samples is conducted using H$_2$ and CO + H$_2$O (15). The greatest difference between the two (while it is the same fraction under almost the same conversion) in a CO + H$_2$O system, there is a scarcity of oxygen containing structure. Further when H$_2$ is used, in asphaltene the σal' portion (hydrogen bonded α carbon from aromatic carbon) is smaller than that produced in the CO + H$_2$O system. Still further under the present conditions, the main reaction involved in molecular lowering is depolymerization, so the nacent hydrogen coming from CO + H$_2$O has a selectivity to attack the ether linkage; in addition it may be considered that H$_2$ relative to this, has a definite activity for the cleavage of the CH$_2$ bridge.

Conclusion

The structural parameter changes of products of coal reduction under various reducing reaction conditions were followed up, and a discussion of the reaction mechanisms involved was made and the following conclusion were obtained.

1) It may be considered that in the hydrogenation reaction of coal, the coal is subjected to consecutive changes in components and reactivity which results in a consecutive molecular lowering.
2) Regarding the chemical reaction observed under comparatively mild hydrogenation, cleavage of linkage between structural units, saturation of aromatic rings, ring opening, dealkylation, deoxygen, desulfurization were seen.
3) When high pressure hydrogen was used as the reducing agent, it could be considered that as a result the addition of hydrogen to α-carbon from the aromatic ring was highest. This was further promoted by increasing the reaction pressure
4) It may be considered that the nacent hydrogen more selectively contributed to the cleavage of the ether bridge and that H$_2$ was more selective than CO + H$_2$O regarding the cleavage of the CH$_2$ bridge.

References

1. Brown, J.K.; Ladner, W.R.; Fuel, 1960, 39, 87
2. Maekawa, Y.; Shimokawa, K.; Ishii, T.; Takeya, G.; Nenryo Kyokai-shi (J. Fuel Soc. Japan), 1967, 46, 928

3. Maekawa, Y.; Ueda, S.; Yokoyama, S.; Hasegawa, Y.; Nakata, Y.; Yoshida, Y.; Nenryo Kyokai-shi (J. Fuel Soc. Japan), 1974, 53, 987
4. Weller, S.; Pelipetz, M.G.; Friedman, S.; Ind. Eng. Chem., 1951, 43, 1572, 1575
5. Falkum, E.; Glenn, R.A.; Fuel, 1952, 31, 133
6. Ishii, T.; Maekawa, Y.; Takeya, G.; Kagaku Kogaku (Chem. Eng. Japan), 1965, 29, 988
7. Itoh, H.; Yoshino, Y.; Takeya, G.; Maekawa, Y.; Nenryo Kyokai -shi (J. Fuel Soc. Japan), 1972, 51, 1215
8. Yoshida, R.; Maekawa, Y.; Takeya, G.; Nenryo Kyokai-shi (J. Fuel Soc. Japan), 1974, 53, 1011
9. Yoshida, R.; Maekawa, Y.; Takeya, G.; Nenryo Kyokai-shi (J. Fuel Soc. Japan), 1972, 51, 1225
10. Yoshida, R.; Maekawa, Y.; Ishii, T.; Takeya, G.; Fuel, 1976, 55, 341
11. Yoshida, T.; et al, under contribution
12. Yoshida, T.; Yokoyama, S.; Yoshida, R.; Ueda, S.; Maekawa, Y.; Yoshida, Y.; Preprint the 13rd coal science symposium of Fuel Soc. of Japan, 1976, 110
13. Yoshida, T.; Nakata, Y.; Yoshida, R.; Imanari, T.; Preprint the 14th coal science symposium of Fuel Soc. of Japan, 1977, 71
14. Yokoyama, S.; Makabe, M.; Itoh, M.; Takeya, G.; Nenryo Kyokai-shi (J. Fuel Soc. Japan), 1969, 48, 884
15. Ueda, S.; Yokoyama, S.; Nakata, Y.; Hasegawa, Y.; Maekawa, Y.; Yoshida, Y.; Takeya, T.; Nenryo Kyokai-shi (J. Fuel Soc. Japan), 1974, 53, 977

RECEIVED April 30, 1980.

Pericyclic Pathways in the Mechanism of Coal Liquefaction

P. S. VIRK, D. H. BASS, C. P. EPPIG, and D. J. EKPENYONG

Department of Chemical Engineering, Massachusetts Institute of Technology, Cambridge, MA 02139

The object of this paper is to draw attention to the possible importance of concerted molecular reactions, of the type
termed pericyclic by Woodward and Hoffman (1), in the mechanism
of coal liquefaction.

In outline of what follows we will begin by brief reference
to previous work on coal liquefaction. The present approach will
then be motivated from considerations of coal structure and hydrogen-donor activity. A theoretical section follows in the form of
a pericyclic hypothesis for the coal liquefaction mechanism, with
focus on the hydrogen transfer step. Experiments suggested by
the theory are then discussed, with presentation of preliminary
results for hydrogen transfer among model substrates as well as
for the liquefaction of an Illinois No. 6 coal to hexane-, ben-
zene-, and pyridine-solubles by selected hydrogen donors.

Previous literature on coal liquefaction is voluminous (2).
Molecular hydrogen at elevated pressures is effective (3) but
more recent work has employed hydrogen donor 'solvents', such as
Tetralin (4), which are capable of liquefying coal at relatively
milder conditions. Donor effectiveness is strongly dependent
upon chemical structure. Thus, among hydrocarbons, several hydroaromatic compounds related to Tetralin ⬡⬡ are known to be
effective (5, 6, 7, 8), while the corresponding fully aromatic
⬡⬡ or fully hydrogenated ⬡⬡ compounds are relatively ineffective (7, 9). Further, among alcohols, isopropanol HO-⟨ and
o-cyclohexyl phenol HO-⟨◯⟩ are effective donors (7) whereas t-
butanol HO-⟨ is not ⟨ (10). These observations have been
theoretically attributed (e.g. 2, 9) to a free-radical mechanism
according to which, during liquefaction, certain weak bonds break
within the coal substrate, forming radical fragments which abstract hydrogen atoms from the donor, thereby becoming stable
compounds of lower molecular weight than the original coal.
According to this free-radical mechanism, therefore, donor

0–8412–0587–6/80/47–139–329$05.00/0
© 1980 American Chemical Society

effectiveness is related to the availability of abstractable hydrogen atoms, and this notion seems to have won general accep-
tance in the literature because it rationalizes the effectiveness
of Tetralin, which possesses benzylic hydrogen atoms, relative to
naphthalene and Decalin, which do not. However, the foregoing
mechanism is less than satisfactory because indane $\bigcirc\hspace{-0.3em}\rangle$, which
has just as many benzylic hydrogens as Tetralin, is relatively
ineffective (7) as a donor. Also, the activity of the alcohol
donors, like isopropanol, would have to involve abstraction of
hydrogen atoms bonded either to oxygen or to an sp^3-hybridized
carbon; this seems unlikely because both of these bonds are relatively strong compared to any that the hydrogen might form with a
coal fragment radical.

The present approach to the coal liquefaction mechanism
evolved from contemporary knowledge of coal structure (e.g. 11,
12, 13), which emphasizes the existence of hetero-atom-containing
structures comprising 2- to 4-ring fused aromatic nuclei joined
by short methylene bridges. From this it is apparent that sigma
bonds between sp^3-hybridized atoms in coal are seldom more than one
bond removed from either a pi-electron system or a hetero-atom
containing substituent. Such a molecular topology is favorable
for pericyclic reactions, which are most prone to occur on skeletons with proximal Π- and σ-bonds activated by substituents. We
therefore hypothesize that the overall interaction between the
coal substrate and hydrogen donor, which eventually leads to
liquefaction, involves a sequence of concerted, pericyclic steps,
which will be indicated in the next section. The novelty of this
approach is twofold; first, the mechanistic concept is essentially
different from any that has hitherto been proposed in coal-related literature and second, it lends itself to quantitative
tests and predictions since the pericyclic reactions envisioned
must obey the Woodward-Hoffman rules (1) for the conservation of
orbital symmetry. It should also be pointed out that if the present approach proves valid, then it will have the engineering
significance that the large volume of recently developed pericyclic reaction theory (14, 15, 16) could be applied to the practical problem of defining and improving actual coal liquefaction
processes.

Theory

In delineating a coal liquefaction mechanism we distinguish
three basic steps, namely rearrangements, hydrogen-transfer, and
fragmentation, all of which are hypothesized to occur via thermally-allowed pericyclic reactions. Some typical reactions appropriate for each step are indicated below using Woodward-
Hoffman (1) terminology:

Rearrangements: Sigmatropic shifts, electrocyclic reactions
Hydrogen-Transfer: Group-transfers
Fragmentation: Group-transfers, retro-ene, cyclo-reversions.

An illustration of how the overall pericyclic mechanism might apply to the decomposition of 1,2 diphenylethane, a model substrate, in the presence of Δ^1-dihydronaphthalene, a model hydrogen-donor, has recently been given (17). In the present work, attention is focussed on the hydrogen-transfer step.

Hypothesis.

The transfer of hydrogen from donors to the coal substrate during liquefaction occurs by concerted pericyclic reactions of the type termed 'group transfers' by Woodward and Hoffman (1).

Consequences.

(A) Allowedness

The Woodward-Hoffman rules (1) state that: "A ground state pericyclic change is symmetry allowed when the total number of (4q + 2) suprafacial and (4r) antarafacial components is odd".

To illustrate how this applies in the present circumstances we consider a possible group transfer reaction between Δ^2-dihydronaphthalene, , a hydrogen donor, and phenanthrene, , a substrate (hydrogen acceptor) which models a polynuclear aromatic moiety commonly found in coal. In the overall group transfer reaction:

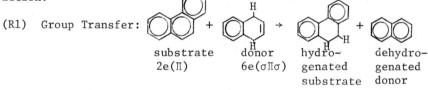

(R1) Group Transfer:

substrate   donor   hydro-   dehydro-
2e(Π)   6e(σΠσ)   genated   genated
                               substrate   donor

hydrogen is transferred from Δ^2-dialin to the phenanthrene, producing 9,10-dihydrophenanthrene and naphthalene; this reaction is slightly exothermic, with $\Delta H_r^\circ \sim -8$ kcal as written. The electronic components involved are 2e(Π) on the substrate and 6e(σΠσ) on the donor and from the Woodward-Hoffman rules it can be seen that the reaction will be thermally forbidden in either the supra-supra stereochemistry (which is sterically most favorable) or the antara-antara stereochemistry (sterically the most unfavorable) but will be thermally allowed in either the antara-supra or supra-antara modes, both of the latter being possible, but sterically difficult, stereochemistries. A reaction profile for (R1) is sketched in Figure 1, showing energy versus reaction coordinate as well as transition state stereochemistry for both the forbidden supra-supra and the allowed antara-supra modes. We note next that changing the donor from Δ^2-dialin to Δ^1-dialin , changes the donor electronically from a 6e(σΠσ) component to a 4e(σσ) component, of contrary orbital symmetry. Thus for the overall reaction:

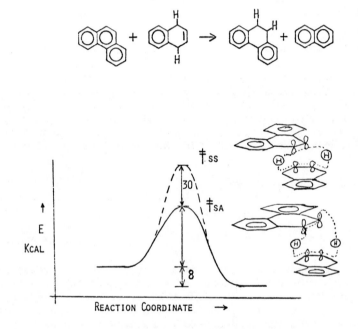

Figure 1.   Schematic reaction profile for group transfer

(R2) Group Transfer: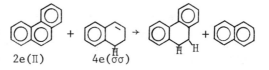

$$2e(\pi) \qquad 4e(\bar{\sigma}\sigma)$$

the supra-supra (and antara-antara) modes will be thermally allowed while the antara-supra (and supra-antara) modes are forbidden. The practical upshot of this is that the chemically very similar dialin isomers should exhibit strikingly different reactivities in concerted group transfers with a given substrate, on account of their opposite orbital symmetries. Note that no reasonable free-radical mechanism could predict profound differences in the donor capabilities of these dialin isomers. Orbital symmetry arguments can also be extended to differences in the substrates. Thus anthracene ⬡⬡⬡ is a $4e(\pi\pi)$ component, of symmetry opposite to that of phenanthrene ⬡⬡ which is a $2e(\pi)$ component; therefore, with a given donor, such as Δ^2-dialin ⬡⬡, group transfers that are symmetry-forbidden with phenanthrene will be symmetry-allowed with anthracene in the same sterochemistry. Accordingly, the reaction:

(R3) Group Transfer: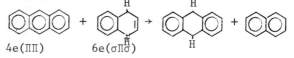

$$4e(\pi\pi) \qquad 6e(\sigma\pi\sigma)$$

is a $4e(\pi\pi) + 6e(\sigma\pi\sigma)$ group transfer, which is thermally-allowed in the supra-supra stereochemistry whereas the analogous reaction (R1) with phenanthrene was forbidden.

Generalization of the preceding suggests that there exist two basic classes of donors (and acceptors), of opposite orbital symmetries, which will respectively engage in group transfer reactions either as (4n+2) electron components or as (4n) electron components. Donors with (4n+2)e will, in general, transfer hydrogen to (4m)e acceptors, the most favorable supra-supra sterochemistry being presumed in each case. Among the (4n+2)e class of hydrogen donors, the first (n=o) member is molecular hydrogen and the second (n=1) member is the but-2-ene moiety, while among the (4n)e class of donors, the first (n=1) member is the ethane moiety, the second (n=2) the hexa-2,4-diene moiety. Among hydrogen acceptors, the (4m+2)e class has the ethylene and hexa-1,3,5-triene moieties as its first two members, while the (4m)e class of acceptors possesses the buta-1,3-diene and octa-1,3,5,7-tetraene moieties as its first two members. Each of the foregoing series can be continued straightforwardly for n>2.

Interestingly, the nature of allowed donor-acceptor interactions suggests that donor class will be conserved in any hydrogen transfer sequence. Thus a (4n+2)e donor, say, will transfer hydrogen to a (4m)e acceptor, and the latter, upon accepting the hydrogen, will evidently become a (4m+2)e donor, of the same class as the original donor.

The donor and acceptor classes illustrated with hydrocarbons can be directly extended to include hetero-atoms. For example, the alcohol moiety $H{\diagdown}O{-}{\diagup}H$ would be a 4e donor, of the same orbital symmetry as the ethane moiety $H{\diagdown}{\diagup}H$. Similarly the carbonyl moiety O= would be a 2e acceptor, analogous to an ethylene moiety = in terms of orbital symmetry. Thus hydrogen transfer reactions of molecules with hetero-atoms should have the same allowedness as reactions of their iso-electronic hydrocarbon analogues.

(B) Reactivity

The actual rates of thermally-allowed pericyclic reactions vary vastly, and frontier-orbital theory (14, 15, 16) has proven to be the primary basis for quantitative understanding and correlation of the factors responsible. It is therefore of interest to find the dominant frontier orbital interactions for the group transfer reactions hypothesized to occur.

The HOMO (highest occupied molecular orbital) and LUMO (lowest unoccupied MO) levels for hydrogen donors used in coal liquefaction are not yet well known, but the principles involved can be illustrated with the group transfer reaction between molecular hydrogen, a (4n+2)e donor with n=0, and naphthalene, a (4m)e acceptor with m=1:

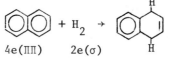

(R4) Group Transfer:

$4e(\pi\pi)$ $2e(\sigma)$

An approximate frontier orbital (FO) interaction diagram for this system is presented in Figure 2. This shows that the dominant FO interaction, i.e., that with the lowest energy gap, is between HOMO(H$_2$)-LUMO (⊙⊙); details of the interaction are given towards the bottom of the figure showing the respective phases and coefficients involved.

The preceding indications regarding the dominant FO interaction in hydrogen-transfer reactions suggests that they would be facilitated by a reduction in the HOMO (hydrogen donor)-LUMO (hydrogen acceptor) energy separation. Thus donor effectiveness should be enhanced by increasing the donor HOMO energy, e.g., by electron-releasing or conjugative substitution, whereas acceptor effectiveness should be enhanced by lowering the acceptor LUMO level, e.g. by electron-withdrawing or conjugative substitution, or by complexation with Lewis acids. As a practical example relevant to our experiments, we should expect Δ^2-dialin to be a more effective (4n+2)e type of hydrogen donor than Tetralin because vinylic substitution in the former should raise the level of the hydrogen-containing HOMO. For the same reason, Δ^1-dialin should also be a more effective (4n)e type of donor than Tetralin . Note that the Δ^1- and

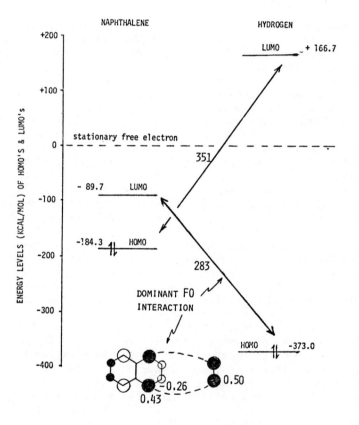

Figure 2. Frontier orbital energy levels for the reaction of hydrogen and naphthalene

Δ^2-dialins have the opposite orbital symmetries; their reactivities cannot therefore be meaningfully compared, inasmuch as their reactions with a common acceptor cannot both be allowed with the same stereochemistry.

Experiments

(A) Model Compounds

From the theoretical discussion it follows that the present hypothesis for the hydrogen transfer mechanism can be tested by a study of reactions between hydrogen donors and acceptors of opposite orbital symmetries. Our experimental grid is shown in Table 1.1. The coal substrate was modelled by anthracene, a (4m) e acceptor, and by phenanthrene, a (4m+2)e acceptor; both of these aromatic C14 moieties exist in coal and are found in coal-derived liquids. The hydrogen donor solvent was modelled by a number of cyclic C10 compounds derived from naphthalene by hydrogenation. Among these, the Δ^2- and Δ^1-dialin isomers were of principal interest, being respectively hydrogen-donors of the (4n+2)e and (4n)e classes; the Tetralin and Decalin served as control solvents, the former being very commonly used in coal liquefaction experiments. For the 2 x 2 matrix of possible hydrogen-transfer reactions between the model C14 substrates and C10 dialin solvents, the Woodward-Hoffman rules predict that reaction in the favorable supra-supra stereochemistry should be thermally-allowed for (Δ^1-dialin + phenanthrene) and (Δ^2-dialin + anthracene) and thermally-forbidden for (Δ^1-dialin + anthracene) and (Δ^2-dialin + phenanthrene). These predictions are shown in Table 1.1 with $\sqrt{}$ denoting the allowed and x the forbidden reactions.

The experiments were conducted batchwise in small stainless-steel pipe-bombs immersed in a molten-salt bath that was maintained at a desired, constant temperature. Pipe-bomb heat-up and quench times, on the order of 1 min each, were negligible compared with reaction times, which were on the order of 1 hr. The reagents used were obtained commercially; all were of purity > 98% except for the Δ^2-dialin which had a composition of (⬡⬡, ⬡⬡, ⬡⬡, ⬡⬡) = 7, 9, 20, 64) mol%. The proportions of substrate to solvent were maintained constant, with the C14 substrates as limiting reactants in all cases. The extent of reaction was measured by proton-NMR spectroscopy on samples of the whole reaction batch, as well as of each of the C10 and C14 fractions separated by vacuum distillation and liquid chromatography.

Experimental results are shown in Table 1.2, which quotes the observed percentage conversion of each of the model C14 substrates to their di-hydro derivatives by each of the model C10 solvents. Consider first the column for the anthracene substrate, showing its conversion to 9,10 dihydroanthracene after 2 hr at 300 C in various solvents. The conversion by ⬡⬡ (58%), is an order of magnitude greater than that by ⬡⬡ (5%), in striking

Table 1.1 Model Compound Experimental Grid

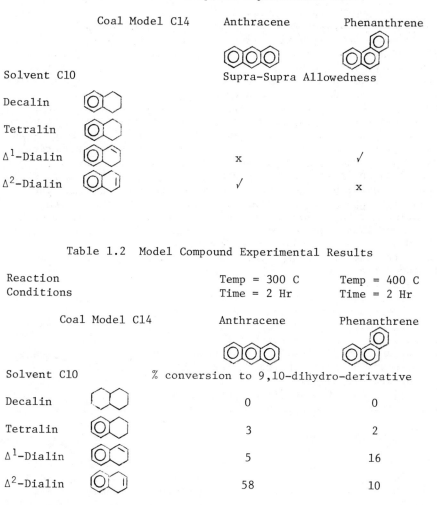

Coal Model C14   Anthracene   Phenanthrene

Solvent C10   Supra-Supra Allowedness

		Anthracene	Phenanthrene
Decalin			
Tetralin			
Δ^1-Dialin		x	✓
Δ^2-Dialin		✓	x

Table 1.2 Model Compound Experimental Results

Reaction Conditions		Temp = 300 C Time = 2 Hr	Temp = 400 C Time = 2 Hr
	Coal Model C14	Anthracene	Phenanthrene
Solvent C10		% conversion to 9,10-dihydro-derivative	
Decalin		0	0
Tetralin		3	2
Δ^1-Dialin		5	16
Δ^2-Dialin		58	10

Note: See text for reagent purities.

accord with the theoretical predictions according to which the reaction with [structure] was allowed while that with [structure] was forbidden. Note too that conversions with the control solvents [structure] (3%) and [structure] (no reaction) were less than those with the test solvents, verifying that the latter were indeed the more active. Reactions with phenanthrene substrate required rather more severe conditions, namely 2 hr at 400 C, than anthracene. While the lower reactivity of phenanthrene relative to anthracene is generally well known, in the present context it can directly be attributed to the phenanthrene possessing the higher energy LUMO. The conversions observed, [structure] (16%) > [structure] (10%), are in accord with theoretical predictions, and appreciably exceed the conversions obtained with the control solvents [structure] (2%) and [structure] (no reaction). It is interesting that the observed selectivity of hydrogen-transfer from the Δ^1- and Δ^2-dialins to phenanthrene, respectively (0.6/1), is not as great as that to anthracene, respectively (12/1). Possible reasons for this are first that whereas anthracene is essentially always constrained to interact with supra-stereochemistry at its 9, 10 positions, the phenanthrene structure admits a possible antara-interaction across its 9, 10 positions and this latter might have permitted a thermally-allowed (antara-supra) hydrogen-transfer from Δ^2-dialin. Second, the Δ^2-dialin used contained some Δ^1-dialin impurity, which could not contribute to anthracene conversion (forbidden) whereas it might have contributed to the phenanthrene conversion (allowed); third, the dialins have a tendency to disproportionate, to naphthalene and Tetralin, at elevated temperatures and this might have influenced the results for phenanthrene, which were at the higher temperature.

In summary, the Δ^1- and Δ^2-dialin isomers have been shown to be appreciably more active than Tetralin (and decalin) in transferring hydrogen to anthracene and phenanthrene. The observed selectivity of this hydrogen transfer is in accord with the Woodward-Hoffman rules for group transfer reactions, anthracene conversions being in the ratio ([structure] / [structure]) = 12/1 >> 1 while phenanthrene conversions are in the ratio ([structure] / [structure]) = 0.6/1 < 1. The quantitative differences in the selectivities observed with anthracene and phenanthrene are being further explored.

(B) Coal Conversion

The dialin donor solvents were also used directly in coal liquefaction studies. Inasmuch as details of coal structure are unknown, the present theory can only be tested in a qualitative way, as follows. First, if the liquefaction of coal occurs under kinetic control with hydrogen-transfer from the donor solvent involved in the rate-determining step, then we should expect the dialin donors to be more effective than the control solvent Tetralin (and also Decalin). This is suggested by the theory because the dialins possess higher energy HOMOs than Tetralin and according to the frontier-orbital analysis given previously, the hydrogen-transfer reactions of the dialins should therefore be kinetically favored relative to those of Tetralin. Second, according to the present theory, donor symmetry is preserved during hydrogen transfer, i.e., a donor of a given class is capable only of interaction with acceptors of the complementary class. Now since the coal substrate likely contains hydrogen acceptors of both $(4m)e$ and $(4m+2)e$ classes, a mixture of solvents containing donors of both $(4n+2)e$ and $(4n)e$ classes should be more effective in hydrogen transfer than a single solvent of either class which could interact with only one of the two possible classes in coal. Thus, in principle, for each coal there should exist an 'optimal' solvent which contains hydrogen donors of opposite symmetries in proportions that are matched to the proportions of the complementary hydrogen acceptors in the coal. For these reasons a mixture of Δ^1- and Δ^2-dialins might be a better donor solvent than either the Δ^1- or the Δ^2-dialin alone, and there may exist an optimal mixture composition that is characteristic of a given coal.

Liquefaction experiments were performed on a sample (18) of Illinois No. 6 coal, from Sesser, Illinois. Proximate and elemental analyses of this high volatile A bituminous coal are given in Table 2. The coal, of particle size 600-1200 microns (32x16 mesh), was dried at 110 C in a nitrogen blanketed oven prior to liquefaction. A solvent to coal weight ratio of 2.0 was used in all experiments, which were conducted in tubing bomb reactors that were immersed in a constant-temperature bath for the desired time while being rocked to agitate the reactor contents. At the end of each experiment, the reactors were quenched and their contents analyzed to ascertain the amounts of each of hexane-, benzene-, and pyridine-solubles (plus gas) produced from the original coal. The procedure used for all analyses is described for the pyridine-solubles (plus gas). First the reactor contents were extracted with pyridine and the residue dried on a pre-weighed ashless filter paper to provide the gravimetric conversion to pyridine-solubles, defined as $100(1-(\text{daf residue/daf coal}))$. Second, the residue was ashed in a furnace for 3 hours at 800 C, yielding the ash-balance conversion to pyridine solubles, defined as $100(1-(a/A))/(1-(a/100))$ where A and a were respectively the weight percentages of ash in the residue and in the original coal. The conversions obtained from each of the two methods normally agreed to within ±2 weight percent and were averaged to provide final values.

Results showing the effectiveness of the Δ^1- and Δ^2-dialins in coal liquefaction relative to control solvents, naphthalene, Decalin, and Tetralin, are presented in Tables 3.1 and 3.2. In both these tables, each row provides the conversion of the coal sample to each of hexane-, benzene-, and pyridine-solubles (plus gases) by the indicated solvent. Table 3.1 contains data derived at a temperature of 400 C and a reaction time of 0.5 hr. Among the control solvents, it can be seen that the naphthalene

Table 2.   Coal Sample Characterization

Origin: Illinois No. 6 from Sesser, Illinois
Rank: Bituminous, High Volatile A

Proximate Analysis:	VM	FC	Ash	Total		
(wt% dry basis)	37.3	56.7	6.0	100.0		

Elemental Analysis:	H	C	N	O	S	Total
(wt% daf)	5.4	82.0	1.6	10.2	0.8	100.0

Table 3.   Coal Liquefaction Results

1.   Temperature = 400 C : Reaction Time = 0.5 hr

Solvent		Coal Conversions to		
		Hexane-	Benzene-	Pyridine-
		Solubles (plus gas), wt%		
Naphthalene		8.0	–	29.7
Decalin		10.8	22.5±2.6	32.5
Tetralin		20.9	43.9±1.4	70.2
Δ^1-Dialin		24.0	39.6±1.9	71.3
Δ^2-Dialin		28.6	44.9±1.5	81.4

2.   Temperature = 300 C : Reaction Time = 0.5 hr

Decalin		0	0	–
Tetralin		0	2.4±0.6	3.7
Δ^2-Dialin		0	9.0±1.5	19.3

and decalin give similar results and are both much less effective
than tetralin, the yields of each of hexane-, benzene-, and pyridine-solubles obtained with the former being roughly half of those
obtained with the latter. The greater effectiveness of Tetralin
as a donor-solvent relative to naphthalene and decalin is in
agreement with previous studies (2, 7, 9). Further, the absolute
conversion to benzene-solubles (plus gases) obtained with tetralin in the present work, namely 44 weight percent, compares fav-
orably with the values of 36 and 47 weight percent reported by
Neavel (9) for comparable HVA and HVB bituminous coals at similar
reaction conditions. The accord between the control solvent
liquefaction data shown in Table 3.1 and the literature permits
us to place some confidence in the present experimental procedures. Turning now to the dialin donors, for which coal liquefaction data have not hitherto been reported, it can be seen from
Table 3.1 that, relative to Tetralin, the Δ^1-dialin yielded appreciably more hexane-solubles, somewhat less benzene-solubles,
and approximately the same pyridine-solubles. Also relative to
tetralin, the Δ^2-dialin yielded appreciably more hexane-solubles, approximately the same benzene-solubles, and appreciably
more pyridine-solubles. It is apparent that the dialins, espec-
cially the Δ^2-dialin, were more effective donor solvents than
tetralin in liquefaction of the present coal sample. While no
precise chemical interpretation can be attached to the quantities
used to measure liquefaction, the pyridine-solubles roughly represent the extent to which the coal substrate is converted,
whereas the hexane-solubles are a measure of the final, oil, product (the benzene-solubles represent an intermediate). Accordingly, from Table 3.1, the Δ^2-dialin increased the coal conversion by 16 percent and product oil formation by 37 percent relative to Tetralin. A few liquefaction experiments were also conducted at a temperature of 300 C and a reaction time of 0.5 hr,
with results reported in Table 3.2. Of the control solvents.
Decalin yielded neither hexane- nor benzene-solubles, while Tetralin yielded no hexane-solbules but did yield the indicated
small amounts of benzene- and pyridine-solubles. The Δ^2-dialin
yielded no hexane-solubles but provided appreciable amounts of
benzene- and pyridine-solubles. The conversions seen in Table
3.2 are all much lower than the corresponding conversions in
Table 3.1, undoubtedly a consequence of the lower reaction temperature, 300 C versus 400 C, reaction times being equal. Finally, Table 3.2 shows that the Δ^2-dialin solvent produced 3 times
the benzene-solubles and 5 times the pyridine-solubles produced
by Tetralin, a striking re-inforcement of the indication from
Table 3.1 that the dialins were the more effective donors.

A second series of liquefaction experiments were conducted
to test the theoretical suggestion that a mixture of Δ^1- plus
Δ^2-dialin isomers might be a more effective solvent than either
one of the dialins alone. Preliminary results at 400 C and 0.5
hr reaction time, are shown in Table 4 which quotes the ratio of

Table 4.   Coal Liquefaction by Mixed Dialin Solvent

Temperature = 400 C : Reaction Time = 0.5 hr

Solvent: 1:1 mixture of Δ^1- and Δ^2-dialins

	Hexane-	Benzene-Solubles	Pyridine-
Conversion Ratio r	1.08	1.01	1.11

Note: r = $p_m/0.5(p_1+p_2)$ where p_m, p_2 are respectively the wt%
       conversions obtained with mixed solvent, Δ^1-dialin and
       Δ^2-dialin.

the conversion of each of hexane-, benzene-, and pyridine-solubles obtained with a solvent mixture containing equal amounts of
Δ^1- and Δ^2-dialins relative to the average of the corresponding
conversions obtained with each of the Δ^1-dialin and Δ^2-dialin
solvents separately. (Generally, if p_{mx} was the conversion to
say, pyridine-solubles obtained with a solvent mixture containing
a fraction x of solvent 1, while p_1 and p_2 were the conversions
respectively obtained with the pure solvents 1 and 2 separately,
then the departure of the ratio $r_x = p_{mx}/(xp_1 + (1-x)p_2)$ from
unity will evidently measure the additional effectiveness of the
solvent mixture relative to the separate pure solvents.)   In
Table 4 it can be seen that the coal conversion to hexane-, ben-
zene-, and pyridine-solubles with the dialin mixture was respectively 8, 1, and 11 percent greater than the average for the separate solvents. Further work is being undertaken using purer
Δ^2-dialin solvent to seek the generality of this result and to
discern an optimum solvent mixture.

Conclusions

At typical coal liquefaction conditions, namely temperatures
from 300 to 400 C and reaction times on the order of 1 hr, hydrogen transfer from model C10 donors, the Δ^1- and Δ^2-dialins, to
model C14 acceptors, anthracene and phenanthrene, occurs in the
sense allowed by the Woodward-Hoffman rules for supra-supra
group transfer reactions. Thus, in the conversion of the C14 substrates to their 9, 10 dihydro derivatives the dialins exhibited
a striking reversal of donor activity, the Δ^1-dialin causing
about twice as much conversion of phenanthrene but only one-tenth
as much conversion of anthracene as did Δ^2-dialin.
The dialins were also found to be more effective donor solvents than Tetralin in the liquefaction of an Illinois No. 6 HVA
bituminous coal. For example, at 400 C and 0.5 hr reaction time,
Δ^2-dialin yielded 16% more pyridine-solubles and 37% more hex-

ane-solubles than Tetralin; at 300 C and 0.5 hr reaction time, the Δ^2-dialin yielded 5 times the pyridine-solubles and 3 times the benzene-solubles yielded by Tetralin.

Finally, a mixture containing equal parts of Δ^1- and Δ^2-dialin was found to be a more effective donor solvent than either of the Δ^1- or Δ^2-dialins separately. At 400 C and 0.5 hr reaction time, the mixture of donors yielded 11% more pyridine-solubles and 8% more hexane-solubles than the average for the separate donors.

The preceding experiments offer preliminary support to our notion that pericyclic pathways might be intimately involved in the mechanism of coal liquefaction. More specifically, the results indicate that pericyclic group transfer reactions constitute a plausible pathway for the transfer of hydrogen from donor solvents to coal during liquefaction.

Abstract

We hypothesize that the mechanism of coal liquefaction might involve three general steps, namely rearrangement, hydrogen transfer, and fragmentation, each proceeding via concerted pericyclic reactions. This mechanism is subject to decisive tests because each step must conform to the Woodward-Hoffman rules. Thus, in the hydrogen transfer reaction, there are predicted to be two distinct classes of hydrogen donors (and acceptors), of opposite orbital symmetries, and reaction should be facile between complementary donor-acceptor classes, either a 4n donor + 4n+2 acceptor or v.v. Within a given donor (or acceptor) class, relative reactivity is governed by frontier orbital interactions between the HOMO of the donor and the LUMO of the acceptor. These predictions were tested by experiments. We used Δ^1- and Δ^2-dialins as hydrogenation solvents and anthracene and phenanthrene as model coal moieties and acquired kinetic data for their respective thermal reactions at temperatures of 200-400 C. By the Woodward-Hoffman rules, the thermal reaction between anthracene and Δ^1-dialin is forbidden whereas that with Δ^2-dialin is allowed. The experimental data, e.g. at 300 C and 2 hr holding time, showed anthracene conversions of 6% with Δ^1-dialin and 58% with Δ^2-dialin (Tetralin yielded negligible anthracene conversion under these conditions). Experiments were also conducted with selected donor solvents in batch liquefaction of coal. The dialin donors, which possess HOMOs of higher energy than Tetralin, were rather more effective; e.g. after 30 min at 400 C, an Illinois No. 6 coal yielded (hexane-solubles, pyridine-solubles) wt% of (21,70) with Tetralin and (29,81) with Δ^2-dialin. The agreement between orbital symmetry predictions and the experimental data offers preliminary support of our hypothesis.

Acknowledgment

This work was supported by the US DOE through the Energy Laboratory at Massachusetts Institute of Technology.

Literature Cited

1. Woodward, R.B., and Hoffman, R. "The Conservation of Orbital Symmetry"; Verlag Chemie GmbH: Weinheim, 1970. Pericyclic reaction terminology defined in this text is used in the present paper.

2. Whitehurst, D.D. "Asphaltenes in Processes Coals"; Annual Report EPRI-AF-480 (1977) and references therein.

3. Bergius, F. German Patents 301,231 and 299,783 (1913).

4. Pott, A. and Broche, H. Glückauf, 1933, 69, 903.

5. Weller, S., Pelipetz, M.G., and Freidman, S. Ind. & Eng. Chem., 1951, 43, 1572.

6. Carlson, C.S., Langer, A.W., Stewart, J., and R.M. Hill. Ind. & Eng. Chem., 1958, 50 (7), 1067.

7. Curran, G.P., Struck, R.T., and Gorin, E. Ind. & Eng. Chem. Proc. Des. & Dev., 1967, 6 (2), 166.

8. Ruberto, R.G., Cronaner, D.C., Jewell, D.M., and Seshadri, K.S. Fuel, 1977, 56, 25.

9. Neavel, R.C., Fuel, 1976, 55, 161.

10. Ross, D.S., and Blessing, J.E. ACS Div. of Fuel Chem. Preprints, 1977, 22 (2), 208.

11. Given, P.M., Fuel, 1960, 39, 147 and 1961, 40, 427.

12. Wiser, W.H., in Wolk, R.M. ACS Div. of Fuel Chem. Preprints, 1975, 20 (2), 116.

13. Wender, I., ACS Div. of Fuel Chem. Preprints, 1975, 20 (4), 16.

14. Fujimoto, M., and Fukui, D. "Intermolecular Interactions and Chemical Reactivity and Reaction Paths", in "Chemical Reactivity and Reaction Paths"; G. Klopman, Ed.; Wiley-Interscience: New York, 1974; pp. 23-54.

15. Klopman, G. "The Generalized Perturbation Theory of Chemical Reactivity and Its Applications", in "Chemical Reactivity and Reaction Paths", G. Klopman, Ed.; Wiley-Interscience: New York, 1974, pp. 55-166

16. Houk, K.N. "Application of Frontier Molecular Orbital Theory to Pericyclic Reactions", in "Pericyclic Reactions", A.P. Marchand & R.E. Lehr, Eds.; Academic Press: New York, 1977, pp. 182-272.

17. Virk, P.S. Fuel, 1979, 58, 149.

18. Our coal sample was obtained from Inland Steel Company Research Laboratories, Chicago, Illinois through the courtesy of Professor J.B. Howard of the Department of Chemical Engineering Massachusetts Institute of Technology, Cambridge, Massachusetts.

RECEIVED May 12, 1980.

An Isotopic Study of the Role of a Donor Solvent in Coal Liquefaction

JOSEPH J. RATTO

Rockwell International, Science Center, 1049 Camino Dos Rios, Thousand Oaks, CA 91360

LASZLO A. HEREDY and RAYMUND P. SKOWRONSKI

Rockwell International, Energy Systems Group, 8900 De Soto Avenue, Canoga Park, CA 91304

Abstract

Fully-deuterated Tetralin was used to study the mechanisms of coal liquefaction. Experiments were conducted with Tetralin-d_{12}, deuterium gas and bituminous coal at 400°C and at 15.2-20.7 MPa. The recovered solvent and solvent-fractionated coal products were analyzed for total deuterium content and for deuterium content in each structural position.

A similar atom per cent of deuterium was found in most of the coal products, while in each soluble product preferential incorporation of deuterium was observed in the α-alkyl position. The amounts of exchange and donation of hydrogen to the coal products were determined. Approximately 35% of the hydrogen transfer resulted from donation to the coal. In the recovered Tetralin, 66% of the incorporated protium (hydrogen from the coal) was found in the H_α position. Indications are that not only hydrogen donation but hydrogen exchange via the α-Tetralinyl radical can have a significant role in quenching the reactive species which form from the thermal cleavage of coal.

Mechanisms of the formation of the decalins in the recovered solvent were based on their isotopic contents. The cis-Decalin-d_{18} had a greater protium content than the recovered Tetralin-d_{12}. This suggested that hydrogen transfer from the coal to the tetralin is involved in the formation of cis-Decalin.

In another experiment, naphthalene-d_8 was used to investigate the chemistry of hydrogen transfer between coal and nondonor solvent at 380°C. An analysis of the recovered naphthalene-d_8 showed that approximately 4% of the hydrogen in the coal and in the naphthalene-d_8 exchanged. Most of the protium incorporated in the naphthalene-d_8 was found in the α-position. The coal products contained approximately 2 wt % chemically-bound napththalene-d_8.

In an earlier presentation, we reported on a deuterium tracer method for investigating the mechanisms of coal liquefaction ([1]).

0–8412–0587–6/80/47–139–347$06.00/0
© 1980 American Chemical Society

The research involved the use of deuterium gas as a tracer to
follow the incorporation of hydrogen into coal. Neither donor
solvent nor catalyst was used in those experiments. The liquefaction product was solvent fractionated, and the fractions were examined for deuterium incorporation in each structural position.
Two significant results of that investigation were that deuterium
incorporation was found to vary with product fraction and that preferential incorporation of deuterium was found in benzylic structural positions. The purpose of this research is to extend the
use of the deuterium tracer method to donor solvent reactions.

   A number of basic studies in the area of donor solvent liquefaction have been reported (2-9). Franz (10) reported on the
interaction of a subbituminous coal with deuterium-labelled tetralin, Cronauer, et al. (11) examined the interaction of deuterium-
labelled tetralin with coal model compounds and Benjamin, et al.
(12) examined the pyrolysis of Tetralin-1-^{13}C and the formation of
tetralin from naphthalene with and without vitrinite and hydrogen.
Other related studies have been conducted on the thermal stability
of Tetralin, 1,2-dihydronaphthalene, cis-Decalin and 2-methylin-
dene (13,14).

   In this investigation, a labelled donor solvent was used to
determine which structural positions in the coal products incorporate deuterium and to investigate the exchange of protium in the
coal with deuterium in the donor solvent. It is important to
understand this fundamental chemistry because a number of pilot
plants use donor solvents (15-17). The yields of liquefied coal
products may be improved through a detailed understanding of the
hydrogenation mechanisms.

   The main part of this research deals with the reaction of
deuterium gas and Tetralin-d_{12} with a bituminous coal. In a separate experiment, naphthalene-d_8 was used for investigating the
chemistry of hydrogen transfer between coal and a nondonor solvent.
In each experiment, the coal products and spent solvent were analyzed for toal deuterium content and for deuterium incorporation
in each structural position.

Experimental

   Materials and Apparatus. High volatile A bituminous coal
(80.1% C, 5.1% H, 1.6% N, 3.6% S, 9.6% O, by weight, daf basis,
7.7% ash), -200 mesh, from the Loveridge Mine, Pittsburgh Seam,
was dried in vacuo at 115°C for 4 hours before use in each experiment. Technical grade deuterium (>98 atom % deuterium, <1 ppm
total hydrocarbons) and high-purity nitrogen were utilized. Naph-
thalene-d_8 was purchased from the Aldrich Chemical Co., and tet-
ralin-d_{12} was prepared in our laboratories (18). The isotopic
purities of tetralin-d_{12} and naphthalene-d_8 were determined by
nuclear magnetic resonance (NMR) using p-dioxane as an internal
reference compound. Batch experiments were performed using a 1-
liter stirred autoclave (Autoclave Engineers) or a 0.25-liter

rocking autoclave (Parr).

Experimental Procedure. In a typical liquefaction experiment, the autoclave was charged at room temperature with Tetralin-d_{12}, coal and deuterium gas. In E10, a rocking autoclave was used. In E19, a stirred autoclave was used with a special stirrer which conformed to the shape of the autoclave liner.

Stirring was initiated, and the autoclave was heated to 400°C which required 90 minutes for E10 and 100 minutes for E19. The temperature was maintained at 400°C for 1 hour, then lowered to room temperature. The cooling duration to 300°C was 5 minutes for E10 and 40 minutes for E19. Stirring was terminated at room temperature. Gaseous products were removed for analysis by gas chromatography coupled with mass spectrometry (GC-MS). The reaction products were distilled at reduced pressure to remove the spent donor solvent mixture, and the remaining coal products were solvent fractionated.

The naphthalene extraction experiment was carried out under similar conditions except that nitrogen was used as cover gas instead of deuterium. The spent naphthalene-d_8 was separated from the residue by distillation at reduced pressure. The residue was solvent fractionated with tetrahydrofuran (J. T. Baker Chemical Co.).

Product Analyses. The spent solvent mixture was distilled from the coal products, separated by GC and analyzed by NMR. Samples were also analyzed by GC-MS, Shrader Analytical Labs., Inc., using a Pye-Unicam Model 105 Chromatograph equipped with a flame ionization detector. A 7'x 1/4" OD glass column with 2% OV-17 was used, and the column was temperature-programmed from 100-150°C at 4°C/min. The GC was interfaced to an AEI Model MS-30 Mass Spectrometer operating at maximum sensitivity. The solid and liquid coal products were solvent-fractionated into oil (hexane soluble, HS), asphaltene (benzene soluble, BS), benzene-methanol soluble (BMS) and insoluble residue (benzene-methanol insoluble, BMI) fractions. Solvent fractionation was performed using three ACS reagent grade solvents: hexane isomer mixture, benzene and methanol. Samples of the fractions were combusted, and the resulting water was analyzed by MS (Shrader Analytical Labs., Inc.) to determine the deuterium and protium atom % compositions.

Proton NMR and deuteron NMR spectra of soluble fractions and spent solvent mixtures were obtained by using a JEOL FX60Q FT NMR Spectrometer. A flip angle of 45° was used which corresponds to 14 μs for ^{1}H and 75 μs for ^{2}H. The pulse repetition times were 6.0 and 9.0 s, respectively. Chloroform-d was used as the ^{1}H NMR solvent, and chloroform was used as the ^{2}H NMR solvent.

Results and Discussion

Product Yields and Compositions. The results of two donor

solvent hydrogenation experiments and a coal extraction experiment
are presented in this paper. The experimental conditions and product yields of the three experiments are summarized in Table I.
The results of the two donor solvent experiments (E10 and E19) are
discussed below. The coal extraction experiment (E20) is discussed in a subsequent section.

The donor solvent experiments were conducted at $400°C$ with a
1:1 coal-to-solvent weight ratio; however, different types and
sizes of autoclaves and operating pressures were used. Examination of the weight % yields of E10 and E19 showed that the same
weight % of oil was formed in each experiment, although E10 had a
much higher percentage of total soluble product yield (HS + BS +
BMS). Under the conditions used in E19, a sizeable fraction of
the tetralin was being hydrogenated and cracked rather than transferring hydrogen to stabilize low molecular weight products formed
from radicals generated from the coal. With less hydrogenation of
the radical species, the radicals may have polymerized which would
decrease the yield of the soluble products.

The atom % ^2H values of the solvent-fractionated products are
also shown in Table I. In previous hydrogenation experiments conducted without the use of a donor solvent (1,18), deuterium incorporation increased from the most soluble oil fraction to the insoluble residue. In E10 and E19, contact of the coal with Tetralin resulted in a uniform incorporation of deuterium in almost all
of the four product fractions. In E19, the BMI fraction's high
value of 61 atom % ^2H may be due to direct gas-phase exchange and
deuteration.

Hydrogen Exchange and Addition Mechanisms. A number of different types of reactions can take place during donor solvent hydrogenation. Tetralin can donate deuterium atoms to the coal and
can exchange its deuterium with protium in the coal. Tetralin can
be hydrogenated to form decalins, rearranged to form methylindan
or hydrocracked to form n-butylbenzene. Figure 1 is a summary of
the reaction pathways of Tetralin which have been identified in
our research.

In the interaction of the donor solvent and hydrogen with the
coal, four main reactions can take place. (1) Tetralin can transfer four atoms of hydrogen to the coal to form naphthalene; this
is the donation or addition mechanism. (2) The coal can be hydrogenated directly by gas-phase deuterium. (3) The gas-phase
deuterium can indirectly hydrogenate coal via tetralin-d_{12}. (4)
Tetralin-d_{12} can participate in isotopic exchange of its deuterium
with protium in the coal. Tetralin participates in pathways 1, 3
and 4. The use of deuterium labelling of the donor solvent and gas
allows investigation of reactions 1 and 4. A future experiment is
planned with deuterium and tetralin-h_{12} to investigate pathway 3.

The liquefaction process is initiated by the thermal generation of coal-derived free radicals which in turn react with solvent to form solvent radicals by hydrogen abstraction. These sol-

TABLE I

SUMMARY OF EXPERIMENTAL CONDITIONS AND PRODUCT YIELDS

Experimental Conditions			
Parameter	E10	E19	E20
Cover Gas	$^2H_2$	$^2H_2$	N_2
Solvent	Tetralin-d_{12}	Tetralin-d_{12}	naphthalene-d_8 T
Coal Weight (g)	25	25	25
Solvent Weight (g)	25	25	25
Reactor Volume (liter)	0.25	1.0	1.0
Stirring Rate or Rocking Rate	100 osc/min	100 rpm	100 rpm
Reaction Time (h)	1.0	1.0	1.0
Cold Pressure $^2H_2$ or N_2 (MPa)	8.3	6.9	6.9
Operating Pressure (MPa)	20.7	15.2	15.2
Temperature (°C)	400	400	380

Solvent-Fractionation Products						
	Weight %			Atom Fraction of ^2H (F_y)		
Products	E10	E19	E20	E10	E19	E20
Oil (HS)	16	17	6*	0.38	0.52	0.23*
Asphaltene (BS)	32	10	94**	0.45	0.43	0.06**
Benzene-methanol Soluble (BMS)	8	2		0.35	0.38	
Benzene-methanol Insoluble (BMI)	44	71		0.37	0.61	

* THF soluble.
** THF insoluble.

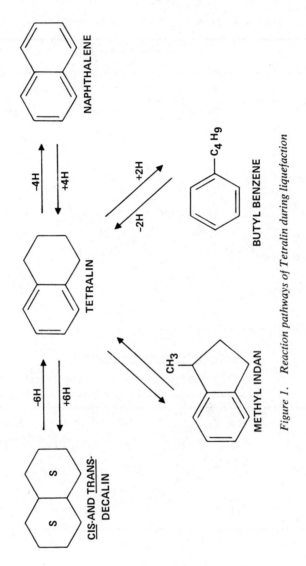

Figure 1.   Reaction pathways of Tetralin during liquefaction

vent radicals can accept or donate hydrogen atoms; therefore, each reversible step shown in Figure 1 can be divided into a number of stepwise additions or removals of hydrogen atoms. The initial stage of dehydrogenation of Tetralin to naphthalene is expanded in Figure 2 to include pathways of hydrogen atom donation and exchange. Tetralin can lose a hydrogen atom to form an α-radical, a β-radical or either of two aromatic radicals. The pathways of donation of hydrogen by Tetralin require the abstraction of α- and β-hydrogen atoms by coal-derived radicals; reactions 1, 2A and 2B in Figure 2 show the main pathways of the donation reaction. Another hydrogen atom lost from the α- or β-Tetralinyl radicals forms either 1,2-dihydronaphthalene by reaction 1 or 1,2- and 1,4-dihydronaphthalene by reactions 2A and 2B. Loss of two more alkyl hydrogens forms naphthalene. The dihydronaphthalenes, which have been detected in low concentrations (6), were not observed under our experimental conditions.

Exchange takes place when a deuterium atom is abstracted from one of the different positions of Tetralin-d_{12} to form a solvent radical, and this radical abstracts a protium atom from coal to regenerate Tetralin. This results in a protium enrichment of the tetralin-d_{12}. Enrichment of protium in the aromatic position can occur by reaction 3 in Figure 2 and also by another mechanism. Tetralin, which has protium incorporated by exchange through participation of the α- and β-alkyl radicals, can lose deuterium to form naphthalene. Either ring of the naphthalene has equal probability for deuterium uptake to regenerate Tetralin. In this manner, a protium atom which was incorporated by exchange originally into an aliphatic position could end up in an aromatic position. However, this mechanism has been shown not to occur at 400°C (12). Only 10% of the incorporated protium was found in the aromatic positions; therefore, these reactions are most likely minor exchange pathways under the conditions used in this research.

The ^{1}H NMR spectrum of the spent solvent from E10 is shown in Figure 3. Tetralin and naphthalene absorption peaks are evident in this spectrum, and the peaks at positions 1, 2 and 3 are due to decalins and in part to methylindan and n-butylbenzene. Methylindan and n-butylbenzene were detected and analyzed by GC-MS. In Figure 3, the large difference in amplitude between the H_{ar}, H_{α} and H_{β} absorptions of Tetralin show that protium was incorporated to a greater degree into the H_{α} position than into the other positions. The spectrum also shows that the H_{α} absorption of the naphthalene in the spent solvent is much more intense than the H_{β} absorption.

To determine the principal pathway of exchange, isotopic exchange by functional position was examined. The lower halves of Tables II and III show the amounts of incorporation of protium into the spent Tetralin. In E10, 66% was in the H_{α} position, 23% in the H_{β} position, and 11% in the H_{ar} position. A similar distribution was found in E19. This preferential incorporation in the H_{α} position of Tetralin strongly suggests that the predominant path-

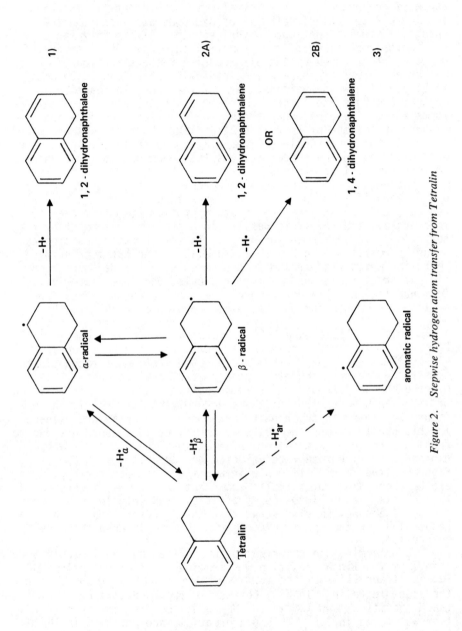

Figure 2. Stepwise hydrogen atom transfer from Tetralin

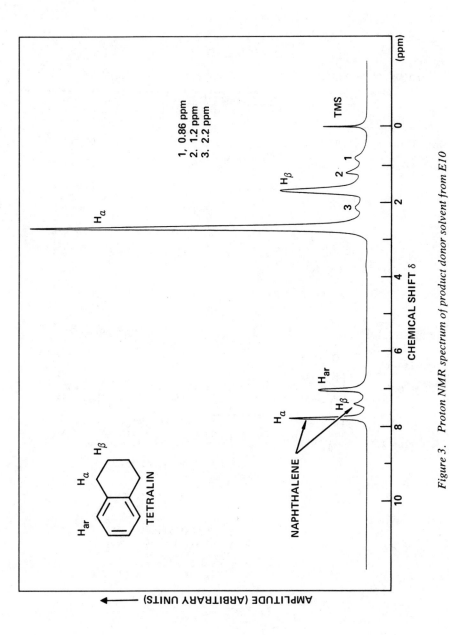

Figure 3. Proton NMR spectrum of product donor solvent from E10

TABLE II

MOLAR COMPOSITION AND ISOTOPIC DISTRIBUTION OF DONOR SOLVENT FROM E10

Molar Composition

Starting Solvent		Spent Solvent	
Tetralin	94 mole %	Tetralin	64 mole %
Naphthalene	6 mole %	Naphthalene	28 mole %
		trans-Decalin	1 mole %
		cis-Decalin	1 mole %
		Methylindan	3 mole %
		Butylbenzene	3 mole %

Isotopic Distribution

Starting Solvent

Tetralin-d_{12}
2.8 atom % ^{1}H
$\begin{cases} H_\alpha & 39\% \ ^1H \\ H_\beta & 14\% \ ^1H \\ H_{ar} & 47\% \ ^1H \end{cases}$

Naphthalene-d_8
<1.0 atom % ^{1}H
$\begin{cases} H_\alpha & 56\% \ ^1H \\ H_\beta & 44\% \ ^1H \end{cases}$

Spent Solvent

Tetralin-d_{12}
21.3 atom % ^{1}H
$\begin{cases} H_\alpha & 66\% \ ^1H \\ H_\beta & 23\% \ ^1H \\ H_{ar} & 11\% \ ^1H \end{cases}$

Naphthalene-d_8
18.2 atom % ^{1}H
$\begin{cases} H_\alpha & 78\% \ ^1H \\ H_\beta & 22\% \ ^1H \end{cases}$

trans-Decalin-d_{18}
14.6 atom % ^{1}H

cis-Decalin-d_{18}
34.0 atom % ^{1}H

Methylindan-d_{12}
41.4 atom % ^{1}H

Butylbenzene-d_{14}
24.7 atom % ^{1}H

TABLE III

MOLAR COMPOSITION AND ISOTOPIC DISTRIBUTION OF DONOR SOLVENT FROM E19

Molar Composition

Starting Solvent		Spent Solvent	
Tetralin	92 mole %	Tetralin	58 mole %
Naphthalene	5 mole %	Naphthalene	22 mole %
trans-Decalin	2 mole %	trans-Decalin	4 mole %
cis-Decalin	1 mole %	cis-Decalin	7 mole %
		Methylindan	5 mole %
		Butylbenzene	4 mole %

Isotopic Distribution

Starting Solvent

Tetralin-d_{12}
<1.0 atom % ^1H
$\begin{cases} H_\alpha & 25\% \ ^1H \\ H_\beta & 38\% \ ^1H \\ H_{ar} & 37\% \ ^1H \end{cases}$

Naphthalene-d_8
<1.0 atom % ^1H
$\begin{cases} H_\alpha & 50\% \ ^1H \\ H_\beta & 50\% \ ^1H \end{cases}$

Decalins-d_{18}
<1.0 atom % ^1H

Spent Solvent

Tetralin-d_{12}
12.3 atom % ^1H
$\begin{cases} H_\alpha & 64\% \ ^1H \\ H_\beta & 24\% \ ^1H \\ H_{ar} & 12\% \ ^1H \end{cases}$

Naphthalene-d_8
10.7 atom % ^1H
$\begin{cases} H_\alpha & 65\% \ ^1H \\ H_\beta & 35\% \ ^1H \end{cases}$

trans-Decalin-d_{18}
7.2 atom % ^1H

cis-Decalin-d_{18}
18.8 atom % ^1H

Methylindan-d_{12}
13.2 atom % ^1H

Butylbenzene-d_{14}
9.9 atom % ^1H

way of hydrogen exchange is through the α-radical. This requires less energy than formation of the β-radical because of resonance stabilization of the α-radical. It should be noted that the α-radical can interconvert to the β-radical by a 1,2-hydrogen atom shift. This reaction would scramble the hydrogen isotopes between the two positions. Since the isotopic compositions of the α- and β-positions were significantly different, it can be concluded that the 1,2-hydrogen atom shift has only a minor role.

The amounts of exchange and addition were calculated from a hydrogen isotope mass balance of the coal products, donor solvent and gas phase hydrogen. The starting and product weights of the coal and hydrogen compositions of the coal and coal products are shown in Table IV. From the values in Table IV, the net amount of hydrogen added to the coal, H, is

$$H = n_H - n_{1_H}^o , \tag{1}$$

where n_H equals the amount of hydrogen in the coal products and $n_{1_H}^o$ is the amount of hydrogen in the starting coal. The amount of exchange, E, is given by

$$E = n_{2_H} - H , \tag{2}$$

where n_{2_H} equals the amount of deuterium incorporated in the coal products through addition and exchange. The values of H and E for E10 and E19 are shown in Table IV. The fraction of deuterium which was incorporated by addition is $H/(H + E)$. These values, which are given in Table IV, indicate that on the average 35% of the hydrogen transfer resulted in addition to the coal.

To summarize the exchange and donation mechanisms of Tetralin, pathway 1 in Figure 2 is the predominant pathway of exchange and donation as determined by preferential incorporation of protium into the α-position of Tetralin. Tetralin-d_{12} loses a deuterium atom from the α-position, and a protium atom is incorporated into the α-position. This equilibrium is the exchange pathway. Continued loss of deuterium from the α-Tetralinyl radical eventually leads to naphthalene, and this reaction is the donation pathway. The large extent of protium incorporation into the α-alkyl position of Tetralin strongly suggests that hydrogen exchange via the α-Tetralinyl radical can have a significant role in quenching the reactive radical species which form by thermal cleavage of the coal.

Formation of Minor Products From Solvent. An explanation of the role of a donor solvent would not be complete without examining four other products which were isolated following interaction of the donor solvent with coal. As shown in Figure 1, cis- and trans-Decalins can be formed by the hydrogenation of Tetralin, methylindan can be formed by rearrangement and n-butylbenzene can be formed by hydrogenolysis. Reaction mechanisms leading to the formation of these products have been investigated (12, 13, 14,

TABLE IV

HYDROGEN ADDITION AND EXCHANGE IN E10 AND E19

Experiment	Weight (g)	Hydrogen (moles/100 g coal)				
		$n_1^o{}_H$	n_H	$n_2{}_H$	H	E	$\frac{H}{(H+E)}$
E10 Coal	25.0	4.7					
E10 Coal Products	23.3		5.7	2.3	1.0	1.3	0.4
E19 Coal	25.0	4.7					
E19 Coal Products	25.5		5.5	3.0	0.8	2.2	0.3

20). The use of a deuterium tracer in this research makes it possible to obtain new information regarding the reactions involved
in the formation of these compounds. Since reactions of this type
result in a loss of hydrogen donor capability of the solvent, research in this area is important to improve the efficiency of coal
hydroliquefaction and the recycleability of the solvent.

In the spent solvents from E10, Table II, naphthalene and tetralin were the major products, and the four others were minor products which totaled 8 mole %. In E19, Table III shows that 20 mole
% of the four minor products were formed, indicating that an appreciable fraction of tetralin was converted to species less effective in the donor process. Protium from the coal, deuterium from
the gas phase or deuterium from the Tetralin is needed to form
these products. An examination of the isotopic composition of
each of the four products as shown in the bottom halves of Tables
II and III allows observations to be made about their formation.

Cis- and Trans-Decalins. Investigations (23-27) in the
area of hydrogenation of naphthalene and Tetralin have shown that
in most cases cis-Decalin forms preferentially when a heterogeneous catalyst is used to catalyze the reaction at low temperature,
while trans-Decalin forms on some catalysts in greater than 50%
yield at high temperatures. Cis addition to a double bond is the
result of the reaction of adsorbed hydrogen atoms with chemisorbed
unsaturated hydrocarbons. The reason for the greater yield of
trans-Decalin at higher temperature is that the partially hydrogenated intermediates can turn over on the catalyst surface resulting in trans addition to the ring junction. There are also other
possible mechanisms such as double-bond migration before cis addition of hydrogen and 1,4-hydrogen addition to 3,4,5,6,7,8-hexahy-
dronaphthalene followed by cis-addition of hydrogen which can give
rise to trans-Decalin. The ratio of cis- to trans-Decalin in E10
was 1.0, while that of E19 was 1.8. The greater ratio of the cis
isomer produced in E19 suggests that the differences in operating
conditions may have influenced the relative formation of the isomers. The better stirring mechanism used in E19 may have caused
Tetralin to have had greater contact with the coal or mineral
matter which can act as a catalyst to favor formation of the cis
isomer. The hypothesis of greater contact with a catalyst in E19
then in E10 is also supported by the fact that four times as much
Decalin was produced in E19 than in E10.

The isotopic composition of the Decalins in the spent solvents
is indicative of the mechanisms of their formation. A total of 6
hydrogen atoms are added to the Tetralin to form Decalin. In E10,
during the course of the experiment, the average protium concentration in the Tetralin increased from 2.3 to 21.3 atom % ^{1}H. The
protium content of 34 atom % ^{1}H in the cis-Decalin was greater
than the protium content in the Tetralin. Likewise in E19, the
protium content in the Tetralin increased from 1.0 to 12.3 atom %
^{1}H, while the cis contained 18.8 atom % ^{1}H. This increase of pro-

tium in the cis-decalins strongly suggests that the coal transfers its hydrogen to Tetralin to form protium enriched cis-Decalin. In contrast, the trans-Decalin had approximately the same protium content (E10: 14.6 atom % ¹H) as the average protium content of the Tetralin. Therefore, either trans-Decalin was formed on the surface of the coal with deuterium from the solvent or gas, or it was formed by isomerization from cis-Decalin with accompanying isotopic exchange with a deuterium source. Because the protium content of the cis-Decalins is much greater than the trans-Decalins, cis-trans isomerization is not very important.

Methylindan and Butylbenzene. Methylindan may be formed by structural rearrangement of Tetralin with no net change in its hydrogen content. However, in E10, the protium content of the methylindan, 41.4 atom % ¹H, was much greater than that of the tetralin, while in E19 the protium content of the methylindan, 13.2 atom % ¹H, was only slightly greater than that of the Tetralin.

Butylbenzene is formed by the hydrogenolysis of Tetralin. The isotopic compositions of the spent butylbenzene (E10: 24.7 atom % ¹H E19: 9.9 atom % ¹H) were only slightly different than the isotopic composition of the spent Tetralins. The detailed mechanisms of the formation of these products are not evident from our current analysis of the data.

Coal Products: Isotopic Distribution by Structural Position. Other workers have also investigated deuterium uptake in coal products by structural position. Schweighardt, et al. (26) examined a centrifuged liquid product from a Synthoil run which was heated to 450°C with deuterium gas, Kershaw and Barrass (27) examined the products from the reaction of coal with deuterium gas using SnCl₂ as catalyst, and Franz (10) examined the products from the reaction of a subbituminous coal with Tetralin-1,1-d₂ at 427°C and 500°C.

¹H NMR and ²H NMR spectra of fractionated coal products from E10 and E19 were recorded and analyzed to determine ¹H and ²H composition for each structural position. In our study, $^1H_{x,y}$ and $^2H_{x,y}$ are defined as the fraction of the ¹H and ²H determined from the integrals of the NMR spectra of a given soluble fraction where y equals HS, BS or BMS and x = γ-alkyl, β-alkyl, α-alkyl or aromatic structural positions. The spectral range of the NMR integrations are given in Table V.

If the atom fraction of deuterium in each separated coal product, given in Table I, is defined as F_y, the atom fraction of protium in the same coal product is therefore $1-F_y$. The total amount of protium and deuterium in a given product fraction y is normalized to unity according to Equation 3,

$$\sum_x \left[{}^2H_{x,y}F_y + {}^1H_{x,y}(1-F_y) \right] = 1.0, \qquad (3)$$

TABLE V

^{1}H AND ^{2}H NMR ANALYSES OF PRODUCT[†]

E10 NMR Analysis

Structural Position	Fraction of Protium $(^{1}H_{x,y}(1-F_y))$			Fraction of Deuterium $(^{2}H_{x,y}F_y)$		
	HS	BS	BMS	HS	BS	BMS
γ-Alkyl 0.0-1.0 ppm	0.07	0.06	0.04	0.03	0.04	0.02
β-Alkyl 1.0-1.9 ppm	0.25	0.13	0.13	0.08	0.06	0.06
α-Alkyl* 1.9-4.5 ppm	0.16	0.14	0.17	0.16	0.19	0.14
Aromatic** 4.5-10.0 ppm	0.14	0.22	0.31	0.11	0.16	0.13
Total $(1-F_y)$	0.62	0.55	0.65 F_y	0.38	0.45	0.35

E19 NMR Analysis

Structural Position	Fraction of Protium $(^{1}H_{x,y}(1-F_y))$			Fraction of Deuterium $(^{2}H_{x,y}F_y)$		
	HS	BS	BMS	HS	BS	BMS
γ-Alkyl 0.0-1.0 ppm	0.04	0.12	0.07	0.04	0.04	0.02
β-Alkyl 1.0-1.9 ppm	0.15	0.16	0.09	0.09	0.09	0.07
α-Alkyl* 1.9-4.5 ppm	0.08	0.09	0.19	0.19	0.21	0.17
Aromatic** 4.5-10.0 ppm	0.21	0.20	0.27	0.20	0.09	0.12
Total $(1-F_y)$	0.48	0.57	0.62 F_y	0.52	0.43	0.38

[†] Estimated error ±0.02.
* The α^2-alkyl value is included in the α-alkyl value.
**The phenolic value is included in the aromatic value.

and the values of $^2H_{x,y}F_y$ and $^1H_{x,y}(1-F_y)$ are shown in Table V.
These values are respectively equal to the atom fraction of deuterium and protium in each structural position. Therefore, the amount
of hydrogen in any structural position is the sum of $^2H_{x,y}F_y$ and
$^1H_{x,y}(1-F_y)$. For example, in the γ-alkyl structural position of
the HS fraction of E10, there is 0.07 (7%) protium and 0.03 (3%)
deuterium for a total of 10% hydrogen in the γ-alkyl position. The
data in Table V show that all functional regions of the soluble
product have some degree of deuterium uptake and that the deuterium
is concentrated in the α-alkyl structural position.

The degree of preferential incorporation of deuterium in each
structural position of each product fraction can be calculated as
the fraction of deuterium in that structural position divided by
the fraction of deuterium in that particular soluble coal product.
This fraction of deuterium in position x of the soluble coal product y can be expressed as

$$P_{x,y} = \frac{^2H_{x,y}F_y}{^2H_{x,y}F_y + {^1H_{x,y}}(1-F_y)} \qquad (4)$$

The criterion for preferential incorporation* is when $P_{x,y}$ is
greater than F_y, or

$$P_{x,y}/F_y > 1.0 \qquad (5)$$

Table VI lists the calculated $P_{x,y}/F_y$ values for the three soluble
coal products from E10 and E19. The values range from 0.58 to 1.63.
The α-alkyl regions in E10 and E19 have $P_{x,y} > F_y$ which indicate
that in liquefaction experiments conducted with a donor solvent,
preferential incorporation occurs in the α-alkyl position.

Extraction With a Nondonor Solvent. It was found by several
researchers that, under certain conditions, bituminous coals can
be solubilized by treatment with a nondonor solvent without using
hydrogen gas. Over 90% extract yields were reported by Storch, et
al. (29), when bituminous coals were extracted with phenanthrene
at its atmospheric boiling point (340°C). Heredy and Fugassi (30)
used tritium- and carbon-14-labelled phenanthrene to investigate
the mechanism of phenanthrene extraction. They found that extraction at 340°C involves a significant amount (10-15%) of exchange
of hydrogen between the coal and the phenanthrene. The authors

---

*In previous presentations (1, 26, 28), the values of $^2H_{x,y}/^1H_{x,y}$
were calculated to show preferential incorporation of deuterium
into structural positions. The ratio $^2H_{x,y}/^1H_{x,y}$ can be expressed
in terms of $P_{x,y}$ and F_y by the equation $^2H_{x,y}/^1H_{x,y} = P_{x,y}(1-F_y)/F_y(1-P_{x,y})$.

TABLE VI

RATIO OF FRACTION OF DEUTERIUM ($P_{x,y}$) IN EACH
STRUCTURAL POSITION OF A SOLUBLE PRODUCT TO THE
DEUTERIUM COMPOSITION (F_y) IN THE SOLUBLE PRODUCT

Structural Position	$P_{x,y}/F_y$ Ratio[+]								
		E10				E19		
	HS	BS	BMS		HS	BS	BMS	
γ-Alkyl	0.79	0.89	0.95		0.96	0.58	0.58	
β-Alkyl	0.64	0.70	0.90		0.72	0.84	1.15	
α-Alkyl*	1.32	1.28	1.29		1.35	1.63	1.24	
Aromatic**	1.16	0.94	0.84		0.94	0.72	0.81	
F_y	0.38	0.45	0.35		0.52	0.43	0.38	

[+] Estimated error ±0.04.
* The α^2-value is included in the α-alkyl value.
** The phenolic value is included in the aromatic value.

suggested that the mechanism of hydrogen exchange involved interaction of the phenanthrene with free radicals formed by the thermal
decomposition of coal. However, radioactive labelling was used,
and this technique did not make it possible to determine the specific structural positions in which the exchanged hydrogen atoms
were incorporated. More recently, the extraction of bituminous
coal with naphthalene was investigated by Neavel (6) and by Whitehurst, et al. (8). At 400° C with contact times of a few minutes,
25-50% extract yields could be obtained. With longer contact
times, the extract yields decreased considerably. It was proposed
(8) that the naphthalene aided the high extract yield by reversibly
accepting hydrogen from and donating hydrogen to the coal during
extraction.

The purpose of this experiment was to investigate the extent
and the structural specificity of hydrogen exchange during the
extraction of bituminous coal with naphthalene. Table I includes
the data of an extraction experiment (E20) conducted with naphtha-
lene-d_8 using nitrogen as the cover gas. In the experiment, the
reactants were heated at 380° C for 1 hour at 2200 psi; the same
apparatus was applied as in E19. After the run, the spent solvent
was separated from the coal by distillation, and the coal and solvent were examined for deuterium and protium incorporation.

Table VII summarizes the isotopic composition of the starting
and spent naphthalene-d_8 and coal. The starting naphthalene contained 98.5 atom % ^2H (1.5 atom % ^1H). During the course of the
reaction, the ^1H content of the naphthalene-d_8 increased by 3.9
atom %. Most of the protium was incorporated into the H_α position,
and the ratio of H_α to H_β was 3.5. The coal residue had 7 atom %
^2H content after extraction.

The coal residue was separated into a THF-soluble fraction
and a THF-insoluble residue. The wt % yields and atom % ^2H compositions are given in Table I. The coal residue was 6 wt % soluble in tetrahydrofuran. The soluble fraction had 23 atom % ^2H content. Evaluation of the ^2H NMR data showed that 85 wt % of this
fraction was derived from the coal and that its deuterium content
was 10%. The chemically-bonded naphthalene-d_8 content of the THF-
soluble fraction, estimated from the ^2H NMR data, was about 15 wt
% or approximately 1 wt % of the coal. The insoluble residue had
6 atom % ^2H content. This indicates that the residue contained
approximately 1 wt % chemically-bonded naphthalene which was estimated from the difference in the atom % ^2H content of the insoluble
residue and recovered naphthalene-d_8. This gives a total chemically-bonded naphthalene-d_8 content of approximately 2 wt %. Similar
results were obtained in extraction experiments made with phenanthrene (30), where it was found that 3-7 wt % of the phenanthrene
was chemically linked to the coal product.

In order to explain the large specific incorporation of protium into the H_α position of the spent naphthalene-d_8, the possible
reaction mechanisms of exchange need to be examined. Naphthalene
may accept a ^1H atom from coal to initiate the reaction as shown

TABLE VII

ISOTOPIC COMPOSITION OF HYDROGEN TRANSFER SOLVENT

Starting	Spent
Naphthalene-d$_8$	Naphthalene-d$_8$
1.5 atom % ^{1}H	5.4 atom % ^{1}H
H$_\alpha$ 59%	H$_\alpha$ 78%
H$_\beta$ 41%	H$_\beta$ 22%
Coal	Coal
~100 atom % ^{1}H	93 atom % ^{1}H

in mechanisms 2A and 2B. Either the α or β positions may react as shown. Coal is not shown in this scheme, but it acts to donate protium atoms or accept deuterium atoms.

The following reasoning was used to eliminate the less probable mechanisms shown in Figure 4. A ^1H atom is added to naphthalene to form an α-radical in reaction 1A and a β-radical in reaction 1B. Both are resonance-stabilized radicals. They can lose either a ^2H atom or a ^1H atom to regenerate naphthalene. We have shown a ^2H atom lost to form a protium-enriched product in reactions 1A and 1B. The fact that we observe a fourfold increase of protium in the α-position of spent naphthalene suggests that reaction 1B is faster than reaction 1A and, therefore, is the predominant mechanism.

In mechanisms 2A and 2B, a ^2H atom is abstracted from either the α or the β positions of the naphthalene to form a o-radical. The intermediates shown in reactions 1A and 2A are radicals in which the unpaired electron is in the α-position. Similarly, in 1B and 2B, the unpaired electron is in the β-position. The terms α- and β-radicals (i.e., reactions 1A and 1B) and σ-radicals (i.e., reactions 2A and 2B) are used, respectively, to distinguish radicals in which the unpaired electron is or is not conjugated with the aromatic π-electron system. Formation of σ-radical intermediates would require higher energy than the formation of the radicals in reactions 1A and 1B. Using the same logic as we have for reactions 1A and 1B, the α-incorporation of ^1H into the spent naphthalene suggests that reaction 2A is more likely than reaction 2B.

We are left with mechanisms 1B and 2A, and a choice between them cannot be made without identification of the predominant radical intermediate formed in this reaction. The high energy required for the formation of σ-radicals suggests that reaction 1B is more likely to take place.

Conclusions

Hydrogen addition and exchange reactions between a bituminous coal and a donor solvent or a nondonor solvent were investigated using fully-deutrated Tetralin and naphthalene. In the experiments conducted with coal, Tetralin-d_{12} and deuterium at 400°C, the ratio of hydrogen exchange to addition was on the average 2.0.

Deuterium-labelled Tetralin and deuterium were used to determine incorporation of hydrogen by structural position in the Tetralin and coal products. Approximately two-thirds of the protium incorporated into the Tetralin-d_{12} was found in the α-alkyl position, indicating that hydrogen exchange between the coal and tetralin involves the α-Tetralinyl radical. In the case of the coal products, the terms $^2H_{x,y}F_y$ and $^1H_{x,y}(1-F_y)$ were used to calculate the deuterium and protium contents of each structural position. Preferential incorporation was found in the α-aliphatic position of the soluble coal products, indicating that these positions participated preferentially in hydrogen exchange and addition. The

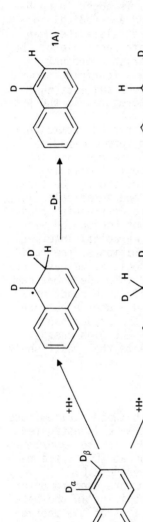
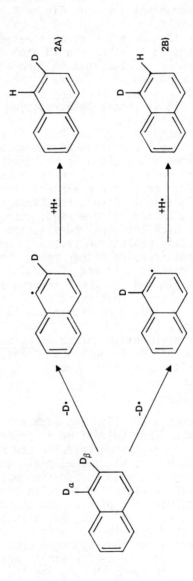

Figure 4. Stepwise hydrogen atom exchange between coal and naphthalene. The starting naphthalene was deuterated completely. To simplify the scheme only α- and β-position are shown.

extent of this preferential incorporation, however, was less than in the case of the α-Tetralinyl position.

An investigation of the isotopic composition of the Decalins, which were formed as minor products in the donor solvent experiments, showed that cis-Decalin was formed preferentially. Its formation and its increased protium incorporation may have resulted from increased contact with the coal surface. Trans-Decalin contained less protium than the Tetralin, which suggests that most of the trans-Decalin was formed with deuterium from the Tetralin-d_{12} and deuterium gas.

When a bituminous coal was extracted with naphthalene-d_8 at 380°C under N_2, 4% of its protium content was exchanged with deuterium in the naphthalene-d_8. Most of the protium was incorporated into the H_α position of naphthalene-d_8, and the ratio of H_α to H_β was 3.5.

Acknowledgement

The authors gratefully acknowledge the technical support of Mr. Tom Johnson. This research was supported by the U.S. Department of Energy under Contract EF-77-C-01-2781.

Literature Cited

1. R. P. Skowronski, L. A. Heredy and J. J. Ratto, Preprints ACS Div. Fuel Chem. 23 (4) 155 (1978).
2. M. Orchin, J. Amer. Chem. Soc. 66 535 (1944).
3. G. P. Curran, R. T. Struck and E. Gorin, Preprints ACS Div. Fuel Chem. 10 (2) C-130 - C-148 (1966).
4. G. R. Pastor, J. M. Angelovich and H. F. Silver, Ind. Eng. Chem., Process Des. and Develop. 9 (4) 609 (1970).
5. J. Guin, A. Tarrer, L. Taylor, Jr., J. Prather and S. Green, Jr., Ind. Eng. Chem., Process Des. Develop. 15 (4) 490 (1970).
6. R. C. Neavel, Fuel 55 237 (1976).
7. I. Schwager and T. F. Yen, Fuel 57 (2) 105 (1978).
8. D. D. Whitehurst, M. Farcasiu, T. O. Mitchell, J. J. Dickert, Jr., "The Nature and Origin of Asphaltenes in Processed Coals", Annual Report, 1977, Mobil Research and Development Corp., Princeton, New Jersey, Report EPRI AF-480.
9. J. E. Schiller and C. L. Knudson, Fuel 57 (1) 36 (1978).
10. J. A. Franz, Fuel 58 (6) 405 (1979).
11. D. C. Cronauer, D. M. Jewell, Y. T. Shah and R. J. Modi, Ind. Eng. Chem. Fundam. 18 (2) 153 (1979).
12. B. M. Benjamin, E. W. Hagaman, V. F. Raaen and C. J. Collins, Fuel 58 (5) 386 (1979).
13. R. J. Hooper, H. A. J. Battaerd and D. G. Evans, Fuel 58 (2) 132 (1979).
14. P. Bredael and T. H. Vinh, Fuel 58 (3) 211 (1979) and P. Bredael and D. Rietvelde, Fuel 58 (3) 215 (1979).

15. C. C. Kang, G. Nongbri and N. Stewart, in "Liquid Fuels From Coal", R. T. Ellington, Ed., Academic Press: New York, 1977, pg. 1.

16 W. P. Epperly and T. W. Taunton, "Exxon Donor Solvent, Coal Liquefaction Process Development", Proceedings of the 13th Intersociety Energy Conversion Engineering Conference, Vol. 1, San Diego, CA, August, 1978, pg. 450.

17. H. E. Lewis, W. H. Weber, G. B. Usnick, W. R. Hollenbach and W. W. Hooks, "Operation of Solvent Refined Coal Pilot Plant at Wilsonville, Alabama", Annual Report, 1977, Catalytic, Inc., Wilsonville, Alabama, Report FE-2270-31.

18. R. P. Skowronski, J. J. Ratto and L. A. Heredy, "Deuterium Tracer Method for Investigating the Chemistry of Coal Liquefaction", Annual Report, 1977, Rockwell International, Canoga Park, CA, Report FE-2328-13, pp. 53-56.

19. R. J. Brunson, Fuel $\underline{58}$ (3) 103 (1979).

20. J. M. L. Penninger and H. W. Slotboom, J. Royal Netherlands Chem. Society $\underline{92}$ (4) 513 (1973).

21. B. B. Corson, Catalysis, Volume III, Chapter 3, 97 (1955).

22. T. I. Taylor, Catalysis, Volume IV, Chaper 5, 257 (1957).

23. R. L. Burwell, Jr., Chem. Reviews $\underline{57}$ 895 (1957).

24. G. C. Bond and P. B. Wells, Adv. in Catalysis $\underline{15}$ 139 (1964).

25. S. J. Thompson and G. Webb, "Heterogeneous Catalysis", John Wiley & Sons, Inc.: New York, 1968, Chapter 6.

26. F. K. Schweighardt, B. C. Bockrath, R. A. Friedel, and H. L. Retcofsky, Anal. Chem. $\underline{48}$ (8) 1255 (1976).

27. J. R. Kershaw and G. Barrass, Fuel $\underline{56}$ (4) 455 (1977).

28. J. J. Ratto, L. A. Heredy and R. P. Skowronski, Preprints ACS Div. Fuel Chem. $\underline{24}$ (2) 155 (1979).

29. C. Golumbic, J. E. Anderson, M. Orchin and H. H. Storch, U.S. Bur. Mines, Rept. Invest. $\underline{4662}$ (1950).

30. L. A. Heredy and P. Fugassi, ACS Advances Chem. Ser. $\underline{55}$ 448 (1966).

RECEIVED May 21, 1980.

Isomerization and Adduction of Hydroaromatic Systems at Conditions of Coal Liquefaction

DONALD C. CRONAUER, DOUGLAS M. JEWELL,
RAJIV J. MODI, and K. S. SESHADRI

Gulf Research and Development Company, P.O. Drawer 2038, Pittsburgh, PA 15230

YATISH T. SHAH

University of Pittsburgh, Department of Chemical and Petroleum Engineering,
Pittsburgh, PA 15213

Fundamental studies of coal liquefaction have shown that the structure of solvent molecules can determine the nature of liquid yields that result at any particular set of reaction conditions. One approach to understanding coal liquefaction chemistry is to use well-defined solvents or to study reactions of solvents with pure compounds which may represent bond-types that are likely present in coal [1,2]. It is postulated that one of the major routes in coal liquefaction is initiation by thermal activation to form free radicals which abstract hydrogen from any readily available source. The solvent may, therefore, function as a direct source of hydrogen (donor), indirect source of hydrogen (hydrogen-transfer agent), or may directly react with the coal (adduction). The actual role of solvent thus becomes a significant parameter.

Our earlier studies [2,3] have measured the reactivity of both hydrocarbon and nonhydrocarbon acceptors with good donor solvents (Tetralin, hydrophenanthrenes), and poor donors (mesitylene). Although the primary role of solvents was observed to be the stabilization of acceptor radicals, appreciable levels of solvent isomerization, polymerization, and adduction also occurred. Herein, these aspects of solvent chemistry have been pursued with the use of ^{13}C labeling techniques to understand the specific reactions.

EXPERIMENTAL

The experimental procedure to carry out the solvent-acceptor reactions have previously been described [2,3]. In summary, the desired amount of solvent was charged to a stirred autoclave and heated to a temperature about 5°C above reaction temperature. The acceptor with additional solvent was injected into the reactor which rapidly came to the desired temperature. The reactor contents were periodically sampled during the run.

0–8412–0587–6/80/47–139–371$05.50/0

Specifically labeled ^{13}C-octahydrophenanthrene and ^{13}C-tetralin were synthesized by Dr. E. J. Eisenbraun. Products from the reactions were analyzed using a combination of the following: (1) GLC using a 100 ft. SCOT capillary column, (2) preparative liquid chromatography using basic alumina [4], (3) preparative HPLC using a column packed with Lichrosorb (silica) and the solvent n-hexane, (4) ^{13}C-NMR using a Varian CFT-20 instrument, and (5) GLC-mass spectra using a duPont 491 instrument.

REACTIVITY OF HYDROAROMATICS

Background

Hydroaromatic compounds are among the most common structures in natural products making up the basic framework of steroids, alkaloids, and mineral oils (petroleum). Hydroaromatic structures are subject to thermal dehydrogenation, unless substituted by gem-dialkyl groups, at bridgehead positions as exhibited by steranes and alkaloids. Dehydrogenation is usually achieved in the presence of catalysts which promote dealkylation, which is a typical precursor to dehydrogenation. Hydroaromatic structures in heterocyclic compounds are frequently more reactive than homocyclics, with respect to dehydrogenation, e.g., tetrahydroquinoline>tetralin and indoline>indan.

Due to the relative ease and reversibility of hydrogenation-dehydrogenation of hydroaromatics, they have been used extensively either as a source or agent for placing hydrogen in hydrogen-deficient species, such as coal. It has frequently been assumed that hydroaromatics in the solvents used for this purpose contain six-membered rings. Little effort has been directed to determining the isomeric forms. It is known that methyl indans are essentially stable to hydrogen-transfer as compared to Tetralin. Due to difficulties in adequately measuring the concentrations of isomeric structures, the above assumption may not be typically valid.

Due to its simple structure and availability, Tetralin is typically used as a model donor solvent for coal liquefaction. For similar reasons, much of the present work was done with Tetralin, as well as octahydrophenanthrene whose structure is believed more allied to true coal-derived recycle solvents [5,6]. Curran et al. [7] observed that Tetralin decomposed to "C_4 benzenes and indan" and that the decomposition seemed to be promoted by coal extracts. They also speculated on several structures for the C_4 benzenes without firm structural evidence. Recent studies by Whitehurst et al. [8] have indicated that Tetralin rearranged to 1-methyl indan and that this rearrangement was solely temperature dependent. These rearrangements were considered reasonably constant at any temperature, proceeded through free radical processes, and were minor side reactions. The present study indicates that isomerization of six-membered hydroaromatics is not limited to tetralin, is quite complex and may be a more serious problem than previously considered.

Reactions of Tetralin and Dihydronaphthalene

Tetralin has been shown to undergo thermal dehydrogenation to naphthalene and rearrangement to methyl indan in either the absence or presence of free radical acceptors [1,2]. The presence of free radical acceptors usually accelerates the rearrangement reaction. Even with alkylated Tetralins, rearrangement still occurs with the formation of di- and tri-alkyl indans.

The basic reactions of Tetralin and derivatives have been extended to the use of 1-^{13}C labels and 1,2-dihydronaphthalene, with and without a source of free radicals. The studies with Tetralin were monitored equally well with ^{13}C-NMR and GLC techniques. The rate constant for the conversion of Tetralin to methyl indan in the presence of dibenzyl at 450°C was 6.4 x 10^{-3} min^{-1} which is consistent with that previously reported [2].

The most significant observation by NMR is the redistribution of the ^{13}C label in the methyl indan isomer. The label is found equally in both methyl and 3-methylene groups as denoted below:

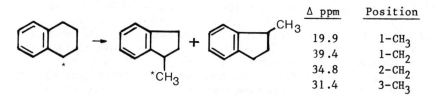

	Δ ppm	Position
	19.9	1-CH$_3$
	39.4	1-CH$_2$
	34.8	2-CH$_2$
	31.4	3-CH$_3$

Concentration of 2-methyl indan and 2-^{13}C-1-methyl indan were very low.

Dihydronaphthalene (DHN) is frequently assumed to be an intermediate in hydrogen transfer reactions. While this appears reasonable, efforts to detect and/or measure this intermediate have never been very successful. Assuming that DHN is present, we have briefly explored its role in hydrogen transfer and methyl indan formation.

Several exploratory experiments were made with unlabeled 1,2-dihydronaphthalene, either neat or with 10% dibenzyl, at 450°C. The runs were made using an agitated 10 cc reactor which was immersed in a preheated sand bath to achieve rapid heating and cooling. It is first noted that the products from experiments at either 15 or 180 minutes contained no unreacted

DHN. Apparently DHN both thermally dehydrogenates to naphthalene and disproportionates to Tetralin and naphthalene. In all of the runs, there was a sizable amount of hydrogen released when the reactors were opened. When DHN was heated at 450°C for either 15 or 180 minutes, the ratio of naphthalene to etralin was 1.8. Increased methyl indan formation occurred with time. With the introduction of dibenzyl, the anticipated [2] increased isomerization of Tetralin to methyl indan occurred. These results suggest that the rearrangement of hydroaromatics does not proceed through the dihydro-intermediate stage, but rather forms directly from the six-membered ring. The dihydro-intermediate only forms during hydrogen transfer.

Reactions of Hydrogenated Phenanthrene

The hydrogenation of phenanthrene proceeds stepwise leading predominantly to the sym-octahydro-stage rather than the asymmetrical form [5]. Either form can function as an excellent donor solvent.

Sym-octahydrophenanthrene (H_8Ph) would be expected to follow the same rearrangement-dehydrogenation reactions as Tetralin, except with more isomer and product possibilities. The reactions shown in Figure 1 illustrate the many structures expected from sym-H_8Ph in the presence of free radical acceptors. Unlike Tetralin, hydrophenanthrenes have multiple structures which each, in turn, form various isomers. The amounts of these isomers are dependent upon the type of hydrogen-transfer reactions and the environment of the system.

A comparison with Tetralin is quite useful, since it indicates the effect that addition of hydroaromatic rings have on the basic problem. Although all the structures shown in Figure 1 are theoretically possible, it is not yet possible to separate each from a total product mixture by current capillary GLC techniques. Our techniques were able to resolve certain groups of compounds which permitted preliminary kinetic calculations. These included mono-iso H_8Ph, di-iso-H_8Ph, iso-H_4Ph, and phenanthrene.

Emphasis in this study was placed upon two reactions carried out at 450°C with sample times between 0 and 180 min. The reference run is that of H_8Ph, neat, and the second run is the hydrogen-transfer reaction of H_8Ph with dibenzyl, in which the benzyl radical is formed at conditions typical of coal liquefaction.

In the presence of dibenzyl, octahydrophenanthrene undergoes both dehydrogenation and isomerization. In this study, we use the kinetic model (refer to Figure 1 for structures):

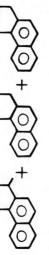
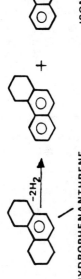

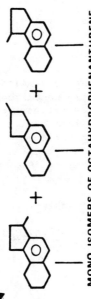
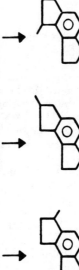
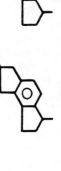
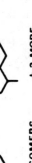
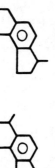

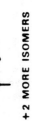

TETRAHYDROPHENANTHRENE

ISOMERS OF TETRAHYDROPHENANTHRENE

OCTAHYDROPHENANTHRENE

$-2H_2$

MONO ISOMERS OF OCTAHYDROPHENANTHRENE

+ 2 MORE ISOMERS

+ 2 MORE

+ 2 MORE

DI - ISOMERS OF OCTAHYDROPHENANTHRENE

Industrial and Engineering Chemistry Fundamentals

Figure 1. Reactions of octahydrophenanthrene (9)

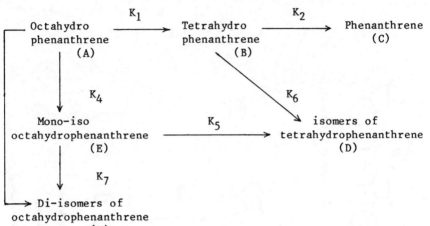

We assume all reactions to be first order and irreversible within the range of the experimental conditions. The governing differential mass balance equations and their solutions have been reported [9]. The values of the constants K_1 through K_6 at 450°C are shown in Table I. A comparison of the experimental data with the theoretical predictions is shown in Figures 2 through 4; the above assumption of a first order reaction appears reasonable.

Table I

KINECTIC CONSTANTS FOR VARIOUS REACTION STEPS

(Reactor Conditions of 450°C, 1500 psig total pressure, and a feed concentration of 30 wt% H_8Ph, 10 wt% dibenzyl, 60 wt% mesitylene)

CONSTANT	(MIN^{-1})
K_1	0.0059
K_2	0.003
K_3	0.0017
K_4	0.0029
K_5	0.0001
K_6	0.002
K_7	0.0035

As shown in Table I, the abstraction of hydrogen is a much more selective reaction compared to isomerization or di-isomerization. Furthermore, isomerization of tetrahydrophenanthrene is a very slow process. As noted in the model, rate constant K_3 is used to denote the direct isomerization of H_8Ph

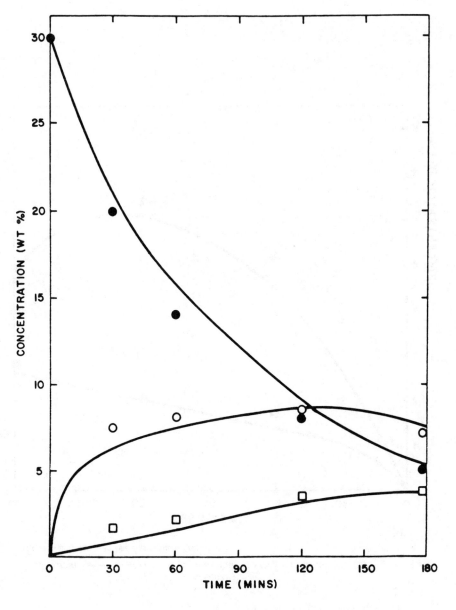

Industrial and Engineering Chemistry Fundamentals

Figure 2.   Concentration of components in the hydrogen abstraction series of reactions in the presence of an acceptor (9): (——), model predicted curve; (●), octahydrophenanthrene; (○), tetrahydrophenanthrene; (□), phenanthrene.

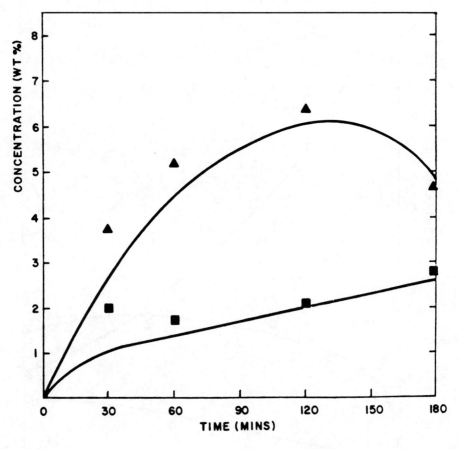

Industrial and Engineering Chemistry Fundamentals

*Figure 3.   Isomerization of solvent to monomethyl isomers (9): (———), model pre-
dicted curve.*

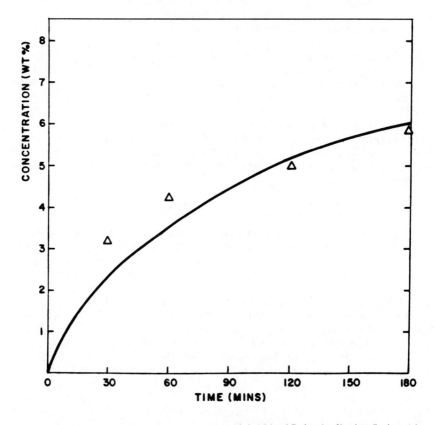

Industrial and Engineering Chemistry Fundamentals

Figure 4. *Isomerization of solvent to dimethyl isomer: (———), model predicted curve (9): (▲), monomethyl (H_8Ph); (■), monomethyl (H_4Ph).*

to di-isooctahydrophenanthrene. No direct evidence exists to prove that both hydroaromatic rings can simultaneously rearrange to the di-isomethyl isomer. However, the kinetic data, namely the plot of concentration versus time, can best be fit with the assumption that this first order reaction occurs.

Studies discussed below with various hydroaromatic systems indicate that these rearrangements proceed through free radical processes as suggested earlier for Tetralins [1-3] and do not require the initial generation of a conjugated olefin group. This approach could generate three-membered ring intermediates similar to those identified by Goodman and Eastman [10] and more recently by Sindler-Kulyk and Laarhoven [11].

It is also noted that the overall rate of isomerization of H_8Ph is about three times that of the isomerization of Tetralin [2]. For reference, the activation energy of the Tetralin isomerization reaction was in the range of 26 to 32 Kcal/g-mole depending upon the presence of a free-radical precursor. Studies have also shown that alkyl Tetralins and recycle solvents containing alkyl groups also rearrange indicating that isomerization is not inhibited by substitution on the hydroaromatic ring [12] However, some steric limitations may exist with a dependence upon the type and size of the attached groups.

STRUCTURAL FEATURES OF HYDROPHENANTHRENES

The presence of hydrophenanthrene isomers was indicated by the observation of numerous GLC peaks with identical parent ions but different fragment ions in their mass spectra. Compounds with methyl substituents always have more intense M^+- 15 ions than those with unsubstituted six-membered rings. Considering the complexity of the total reaction mixtures, liquid chromatography (HPLC) was used to concentrate more discrete solvent fractions for ^{13}C-NMR study.

The spectra of the saturate region of pure sym-H_8Ph, and two monoaromatic concentrates have been observed and the assignment of signals in sym-H_8Ph have been reported [13]. The appearance of new signals at 19 to 21.3 ppm were indicative of methyl groups in a variety of positions on saturated rings. Signals between 30 and 35 ppm were indicative of five-membered rings being formed at the expense of the eight hydroaromatic carbons in the six-membered rings.

The absence of a sharp line at approximately 14 ppm indicated that ring opening to a n-butyl substituent did not occur. Precise mass measurements further showed that each concentrate has the same molecular weight (186) which confirmed that ring opening did not occur as implied by the work of Curran et al. [7] in which experimention was done with Tetralin.

When an acceptor was present (ex., dibenzyl), the solvent products were more complex. The reactions were, therefore, repeated using ^{13}C labeled octahydrophenanthrene (10% [13]C at position 1). The presence of a label not only confirmed the qualitative nature of reactions shown in Figure 1, but provided useful clues as to the real complexity of the structures and pathways for their formation.

A sample of hydrophenanthrenes from a dibenzyl hydrogen-transfer reaction was separated by liquid chromatography into seven fractions. Each fraction (>20 mg) was then analyzed by ^{13}C-NMR, mass spectrometry (70 eV), and ultraviolet spectroscopy and "best fit" structures were then deduced. Minor components were not studied. Pure octahydro- and tetrahydrophenanthrene were used as a "reference base" to compare isomers.

A detailed discussion of these fractions together with the probable structures in each has been presented earlier [9]. The most important observations were (1) every possible position isomer of rearranged octahydro- and tetrahydrophenanthrene were present but not equally distributed; and (2) the benzylic carbons of hydroaromatic rings have migrated to a methyl group (confirmed by ^{13}C label).

A small amount of hexahydrophenanthrene was present in the mixture indicating that the step-wise transfer of hydrogen (loss of two hydrogens) does occur. Three condensed rings apparently provide more stability for this intermediate than does the naphthalene system. The studies with octahydrophenanthrene confirmed that isomerization is not unique to Tetralin. The problem becomes more acute with increasing number of hydroaromatic rings. These structure studies also suggested that tetrahydrophenanthrene may be a key intermediate and should therefore be studied directly in order to understand the effect that condensed aromatics have on the fate of a single hydroaromatic ring.

Studies with Tetrahydroanthracene and Tetrahydrophanthrene

Since the Tetralin studies showed that isomerization yielded predominantly the 1-methyl indan isomer and that the contraction involves the migration of the benzylic carbon to a methyl group, we decided to explore the effect that additional ring condensation has on this chemistry.

1,2,3,4,-Tetrahydroanthracene is the linear benzologue of Tetralin. This compound behaves identical to Tetralin in the presence of dibenzyl with respect to ring contraction, giving a single methyl signal at 19.6 ppm (Figure 5). Contrary to Tetralin, which does not yield a measurable level of 1,2-dihydro-intermediate, one observes the formation of 9,10-dihydroanthrene (36.1 signal). The only other product is anthracene. The yield of 1-methylcyclopentanoanthracene is

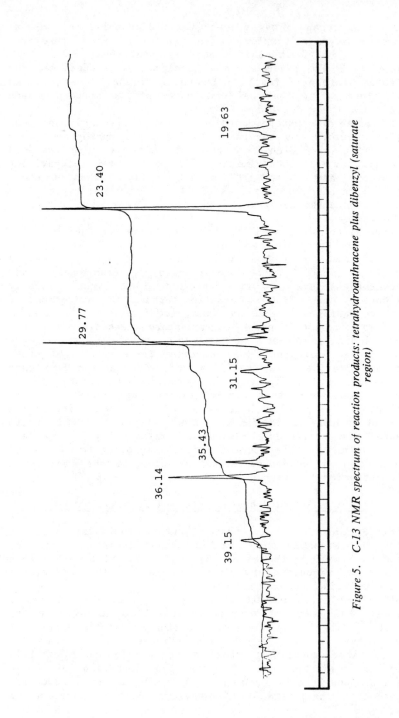

Figure 5.   C-13 NMR spectrum of reaction products: tetrahydroanthracene plus dibenzyl (saturate region)

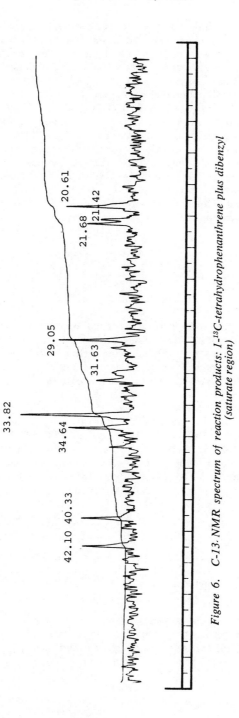

Figure 6. C-13 NMR spectrum of reaction products: 1-^{13}C-tetrahydrophenanthrene plus dibenzyl (saturate region)

slightly less than that in the corresponding Tetralin system.
Anthracene and 9,10-dihydroanthracene are slightly greater than
stoichiometrically predicted by hydrogen transfer. This may
be due to a greater ease of thermal dehydrogenation of
three-ring hydroaromatics.

1,2,3,4-Tetrahydrophenanthrene is the angular benzologue
of Tetralin. Both natural and 10% $1-^{13}C$-enriched H_4Ph were
studied in a manner similar to Tetralin. However, one observes
a much more complex mixture of products with H_4Ph, as seen in
the C-NMR spectrum (Figure 6). Three distinct lines between
20-22 ppm are observed due to the methyl groups on the three
possible isomers from ring contraction. It is also noted that
in the angular system, these signals are at lower field than in
the linear systems (approximately 19.6 ppm). The assignments
of these signals are as follows:

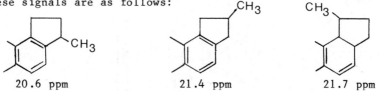

| 20.6 ppm | 21.4 ppm | 21.7 ppm |

A small amount of 9,10-dihydrophenanthrene is observed
(signal at 29.0 ppm) presumably as a result of rapid
isomerization of any 1,2-dihydrophenanthrene formed from the
abstraction of hydrogen from H_4Ph by hydrogen transfer. As is
the case with dihydroanthracene, the 9,10-derivative is the
most stable isomer.

An unexpected observation was the scrambling of labeled
carbon in the H_4Ph solvent during hydrogen transfer
experiments. As determined by NMR, about 25% of the ^{13}C label
at the C-1 position of H_4Ph was found in the C-4 position of
phenanthrene in the product of a three-hour run with dibenzyl
at 450°C. The following nomenclature and shifts in
phenanthrene were used in the determination.

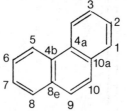

δ_c from TMS	
C-1 and 8	- 128.56 ppm
C-2,3,6,7	- 126.6
C-4 and 5	- 122.6

As shown later, the same signal enhancement of C-4 is seen when
labeled sym- or asym-H_8Ph are used. This type of enrichment at
C-4 implies that concurrent cleavage of the C_1-C_{10a} and C_4-C_{4a}
bonds occur, together with ring inversion.

asym-Octahydrophenanthrene Experiments

Catalytic hydrogenation of phenanthrene to the octahydro-stage produces both sym- and asym-isomers, although the former predominate. Additionally, interconversion of the two forms tends to occur at coal liquefaction conditions. Since the asym- form has not been studied previously, we briefly explored both natural and $1-^{13}C$-enriched asym-H_8Ph with respect to its reactivity with dibenzyl at 450°C.

The spectrum of product of the run with $1-C^{13}$-asym H_8Ph is shown in Figure 7. The GLC results indicate that asym-H_8Ph decomposes to numerous products that are not normally observed with sym-H_8Ph. Based on GLC and NMR spectra, the following observations were made:

1. The condensed cycloparaffin rings crack yielding n-butyl groups, containing, in part, the ^{13}C label in a terminal CH_3 position (signal at 14.0).

2. Ring contraction has occurred to yield at least two methylcylopentane derivatives (signals at 20.6 and 21.7 ppm).

3. Tetrahydro- and dihydrophenanthrenes have been formed (signals at 25.7 and 29.0 ppm).

4. Phenanthrene has been formed (signals at 128.5 and 122.6 ppm).

5. About 25% of the $1-^{13}C$ label has migrated to position C-4 based upon the NMR spectra of phenanthrene formed.

Discussion of Isomerization Results

The studies of rearrangement of hydroaromatics suggest that isomerization is dependent upon the breaking of benzylic carbon bonds. It is promoted by the presence of free radicals. All of the hydroaromatic molecules yield ring contraction products at coal liquefaction conditions. Angular hydroaromatics are much more likely to rearrange to a variety of position isomers than linear hydroaromatics although the rate of isomerization of each separate specie may be different than that of Tetralins.

The above results show that the dihydro-aromatics do not directly contribute to rearrangements. Secondly the dihydro-aromatics rapidly aromatize by hydrogen transfer or disproposionation. This implies that the rearrangement

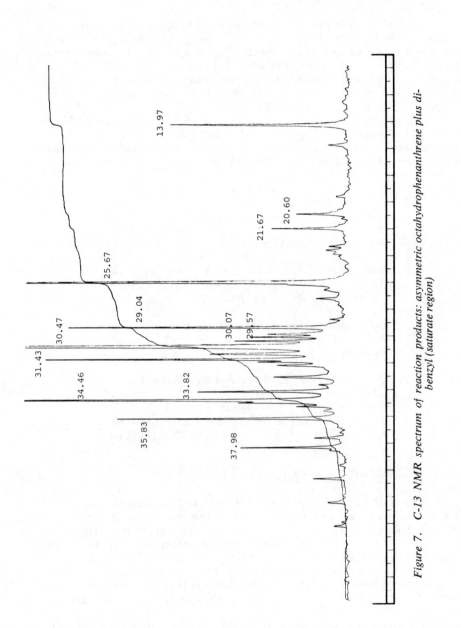

Figure 7. C-13 NMR spectrum of reaction products: asymmetric octahydrophenanthrene plus di-benzyl (saturate region)

proceeds by a free radical process. It is also noted that when free radicals from the acceptor are present, the rate of rearrangement is greatly increased.

SOLVENT ADDUCTION

The primary reaction between good donor solvents, such as Tetralin and octahydrophenanthrene, and acceptors can give rather "ideal" products. For example, at moderate dibenzyl acceptor concentrations (10-20%) dibenzyl is converted only to toluene in these solvents. However, when poor solvents are introduced, secondary reactions become quite important and "non-ideal" products are recovered. The type of secondary products are influenced by the solvent used, the temperature of reaction, and the structure of acceptor molecules. The secondary reactions may well compete with primary reactions to such an extent that kinetics become difficult to model.

As shown earlier [2] mesitylene forms adducts with benzyl radicals concurrent with hydrogen transfer from Tetralin at 450°C. Although not shown in the previous paper, mesityl radicals also formed adducts with Tetralin in mixed systems.

When reactions with oxygen-containing acceptors were performed [3] in the 300-400°C region, the formation of adducts occurred with both Tetralin and mesitylene. This reaction was observed when benzyl radicals were generated from dibenzyl ether, dibenzyl sulfide, benzyl alcohol, and benzaldehyde.

The most surprising observation from low temperature reactions was the formation of adducts between good donor solvents (Tetralin, octahydrophenanthrene, tetrahydroquinoline) and acceptor radicals. The resulting adducts were not of a single predominant structure. In particular, several isomers of toluene-Tetralin were formed as well as di-Tetralin. Several of these reactions were done with D_4-Tetralin which permitted the firm identification of the Tetralin moiety in the adducts. GLC-MS studies indicated that the Tetralin may be bonded to phenyl, benzyl, benzyloxy- or phenoxy-groups, depending on the acceptor used.

Bonding is assumed to be predominately on the hydroaromatic ring since this should be the most reactive site of tetralin donors. This is supported by a large fragment ion at M^+-benzyl. However, based on mesitylene experiments, some bonding on the aromatic ring also occurs.

It appears that the formation of benzyl Tetralin only occurs during the hydrogen transfer reactions at low temperatures (<400°C). If the adduct forms, a certain degree of "depolymerization" could be achieved by isolating the adduct fraction and reacting it with fresh Tetralin at 450°C. Such a sample of adduct from a low temperature (400°C) run with Tetralin and benzyl alcohol was isolated and a second run was made at 450°C with additional Tetralin. Less than 50% of the adduct "depolymerized." The remaining adducts had the same molecular weight but much larger M^+-15 ions indicating the presence of a methyl group. ^{13}C-NMR showed that significant rearrangement had occurred.

Figures 8 and 9 show the partial ^{13}C-NMR spectra of both original and 450°C thermally treated adducts. Each fraction is a mixture of isomers. The lines at 34.2, 38.7, 39.1, and 41.1 are indicative of five-membered rings, while the strong line at 21.2 is indicative of a methyl group on a saturated ring. The following structures would be typical of these signals:

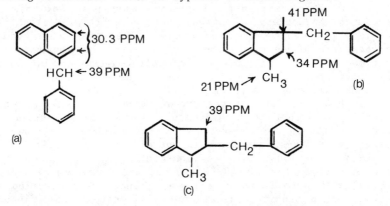

The strong line at 30.0 ppm in both spectra indicate that the six-membered ring is intact and that, when certain positions are substituted by benzyl groups, they resist cleavage. The fact that lines at 38, 39, and 41 ppm disappear upon heating at 450°C suggest that those particular structures are cleaved (unlikely if they are in a five-membered ring) or rearranged to a more stable form. Finally, the observation that benzyl naphthalenes are also present as reaction products confirm that isomer (a) is present and that it can still function as a donor after adduction.

The adduction reactions discussed are not limited to benzyl-type radicals nor to Tetralin solvents. They have been observed with long chain thioether acceptors and donor solvents including dimethyltetralin, octahydrophenanthrene, and tetrahydroquinoline. Donors using an ^{2}H or ^{13}C label have been used to provide further confirmation that the solvent was incorporated in adducts.

CONCLUSIONS

Implications to Coal Liquefaction

Numerous implications on the fundamental chemistry of coal liquefaction can be drawn from the observed reaction of solvent isomerization and adduction. The literature indicates that

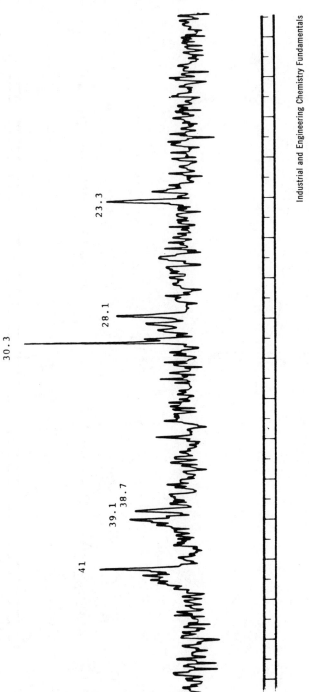

Industrial and Engineering Chemistry Fundamentals

Figure 8. Partial C-13 NMR spectrum of Tetralin adducts formed at low temperatures (9)

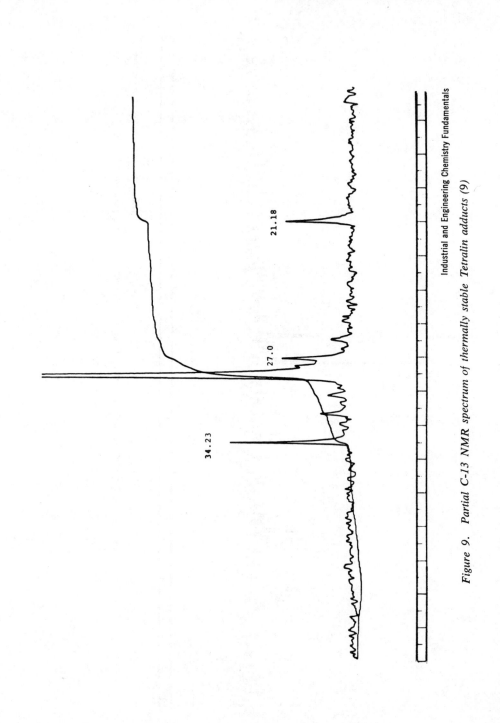

Industrial and Engineering Chemistry Fundamentals

Figure 9.   Partial C-13 NMR spectrum of thermally stable Tetralin adducts (9)

recycle solvents from most coal liquefaction processes consist of 2 to 3 aromatic rings with various degrees of saturation. In this system, high levels of effective hydrogen donors can rearrange to isomers having poor donor quality, as expected from thermodynamic consideration.   The following specific points are noted:

1. The rate of rearrangement of hydroaromatics appears to be first-order with respect to concentration.

2. The rate of isomerization increases with increasing number of hydroaromatic rings. (Previous work [9] has indicated that the rate of rearrangement of hydroaromatics is greatly increased when free radicals are present).

3. A wide range of isomers will be formed from the rearrangement of hydroaromatic compounds.

4. Most hydroaromatic solvents have the capability of becoming irreversibly adducted by acceptor free radicals which could arise from the coal. (On-going work indicates that the presence of oxygen and sulfur functions on the free radicals will enhance adduction.)

ACKNOWLEDGMENT

Funding of this research was provided by the U.S. Department of Energy under Project E(49-18)-2305. We also acknowledge H. Podall and L. Kindley of DOE and A. Bruce King and H. G. McIlvried of Gulf Research for their helpful suggestions, J. C. Suatoni, and T. Best for experimental assistance.

Literature Cited

[1]   Benjamin, B.M., Raaen, V.F., Maupin, P.H., Brown, L.L., and Collins, C.J., Fuel 57, 269 (1978).

[2]   Cronauer, D.C., Jewell, D.M., Shah, Y.T., Kueser, K.A., Ind. Eng. Chem. Fund, 17, 291 (1978).

[3]   Cronauer, D.C., Jewell, D.M., Shah, Y.T., Modi, R.J., Ind. Eng. Chem. Fund 18, 153 (1979).

[4]   Jewell, D.M., Ruberto, R.G., Davis, B.E., Anal. Chem. 44, 2318 (1972).

[5]   Ruberto, R.G., Cronauer, D.C., Jewell, D.M., Seshadri, K.S., Fuel 56, 17 (1977).

[6]   Ibid 25 (1977)

[7]   Curran, G.P., Struck, R.T., Gorin, E., Ind. Eng. Chem. Process Des. Dev. 6 (2), 166 (1967).

[8]   Whitehurst, D.D., Farcasiu, M., Mitchell, T.O., EPRI Annual Report No. AF-480, RP-410-1 (July 1977).

[9]   Cronauer, D.C., Jewell, D.M., Shah, Y.T., Modi, R.I., Seshadri, K.S., Ind. Eng. Chem. Fund. 18, (1979).

[10]   Goodman, A.L. and Eastman, R.H., J.A.C.S. 86, 908 (1967).

[11]   Sindler-Kulyk, M., Laarhoven, W.H., J.A.C.S. 100, 3819 (1978).

[12]   Jewell, D.M., Ruberto, R.G., Seshadri, K.S., "Transferable Hydrogen in Coal Liquids; An Integrated Approach", presented at the 26th Annual Conference on Mass Spectrometry and Allied Topics", A.S.M.S., St. Louis, MO., May 28, 1978.

[13]   Seshadri, K.S., Ruberto, R.G., Jewell, D.M., Malone, H.P., Fuel 57, 111 (1978a).

[14]   Ibid 57, 549 (1978b).

[15]   Mickley, H.S., Sherwood, T.K., and Reed, C.E., Applied Mathematics of Chemical Engineering, 2nd ed., McGraw Hill Book Co., New York, NY (1957).

RECEIVED April 16, 1980.

INDEX

INDEX

395